DEFINITIONS, CONVERSIONS, and CALCULATIONS

for
Occupational Safety and
Health Professionals

Edward W. Finucane, PE, CSP, CIH

LEWIS PUBLISHERS
Boca Raton   Ann Arbor   London   Tokyo

Library of Congress Cataloging-in-Publication Data

Catalog record is available from the Library of Congress.

ISBN 0-87371-863-1

PRINTED IN THE UNITED STATES OF AMERICA
 2 3 4 5 6 7 8 9 0

Printed on acid-free paper

With gratitude,

this work is dedicated to my wife,

Gladys

without whose support and encouragement

it would never have been completed;

and to my teacher,

Professor Andrew J. Galambos

whose teachings and innovations have formed the foundation

upon which my professional consulting business

is based and without which this work

would never have been started.

THE AUTHOR

Edward W. Finucane was born in San Francisco, and raised in Stockton, California. He has earned degrees in engineering from Stanford University, and in business from Golden Gate University. Professionally, Mr. Finucane has been involved in the Occupational Safety and Health field for more than 20 years, during the last twelve of which, he has operated his own professional consulting company, which specializes in this specific area. He is a registered professional engineer, a certified safety professional (comprehensive practice), and a certified industrial hygienist (comprehensive practice). He has had extensive experience in the areas of: ambient gas analysis, gas analyzer calibration, indoor air quality, ventilation, noise and sound, heat and cold stress, and health physics. For the past several years, he has served on the faculty of the *Comprehensive Review of Industrial Hygiene* course offered jointly by the Center for Occupational and Environmental Health (University of California at Berkeley) and the Northern California Section of the American Industrial Hygiene Association.

Table of Contents

Solutions to Example Problems

VENTILATION PROBLEM SOLUTIONS

STATISTICS AND PROBABILITY PROBLEM SOLUTIONS

Preface

This book is intended to serve several purposes:

1. To function as a ready Desk Reference for the Occupational Safety and Health Professional OR Industrial Hygienist. Such an individual, in the normal development of his or her career, will likely have specialized in some relatively specific sub-area of this overall discipline. For such an individual, there will likely be occasions when a professional or job related problem or situation will arise — one that falls within the general domain of Occupational Safety and Health AND/OR Industrial Hygiene, but is outside of this individual's area of principal focus and competence; and is, therefore, not immediately familiar to him or her — for such cases, this Reference Source will hopefully provide a simple path to the answer.

2. To function as a useful Reference Source, Study Guide, or refresher to any individual who is preparing to take either the Core or the Comprehensive Examinations for Certification as an Industrial Hygienist, or as a Safety Professional.

3. Finally, to assist Students who have embarked on a course of study in the Occupational Health or Industrial Hygiene areas. As a fairly concise compilation of all the various important mathematical relationships and definitions that these Students will be called upon to utilize as they progress in their profession, it is hoped that this group, too, may find this work to be of some value.

This book, as a Reference Information Source and Example Problem Workbook, contains virtually every Mathematical Relationship, Formula, Definition, and Conversion Factor that any Professional in this overall discipline will ever need or encounter. Every effort has been made to be certain that the information and relationships in it reflect the very best of the current thinking and technological understanding, as these concepts are currently being used in the field.

The Problem Solution Section of this book contains careful step-by-step Solutions to each of the problems, as well as complete explanations of the reasons and factors that had to be considered and used in completing each of these steps. The underlying goal in generating these very detailed Solutions was that they would constitute a very complete road map that leads from the Problem Statement, itself, all the way to its eventual Solution. The various Problems, having been developed out of the real life professional experiences of the Author, are representative of the actual situations that a professional in these fields will routinely encounter in the normal conduct of his or her profession; because of this, it is hoped that they, too, will be of special value to both the professional and the prospective professional, alike.

Finally, I would like to thank Jon and Brian Lewis of Lewis Publishers for their support and patience; Gladys Finucane for her careful proofreading of all the written copy; William Charney of San Francisco General Hospital, both for his suggestion that a work of this type might well have broader value than I had ever imagined, and for introducing me to Lewis Publishers; and Phillip Finucane for his proofreading and step-by-step checking of the entire Problem Solution Section of this work. In the absence of any one of these basic supporting efforts, I doubt that this work could ever have been completed.

Edward W. Finucane

Definitions of Terms
Worker Safety Related

Threshold Limit Value

The Threshold Limit Value [usually abbreviated, **TLV**] refers to an airborne concentration of some substance of interest, and represents a condition under which it is believed that nearly all workers may be repeatedly exposed, day after day, without adverse effect. This concentration can be, and is, commonly expressed in one of three forms: as an 8-hour Time Weighted Average (**TLV-TWA**); as a Short Term Exposure Limit (**TLV-STEL**); and as a Ceiling Value (**TLV-C**). The overall concept designated by the term or phrase, "Threshold Limit Value", was introduced and promulgated by the American Conference of Government Industrial Hygienists [the ACGIH].

Permissible Exposure Limit

The Permissible Exposure Limit [usually abbreviated, **PEL**] is an airborne concentration of some substance of interest which the Occupational Safety and Health Administration branch [OSHA] of the U. S. Department of Labor [USDOL] had adopted – largely from data furnished earlier by the ACGIH in the development of their listing of Threshold Limit Values. The initial listing of these Permissible Exposure Limits was made in the Z-Tables of Title 29, Code of Federal Regulations, Part 1910.1000, as published in the *Federal Register* on January 19, 1989. Like **TLV**s, **PEL**s are commonly expressed in one of three forms: as an 8-hour Time Weighted Average (**PEL-TWA**); as a Short Term Exposure Limit (**PEL-STEL**); and as a Ceiling Value (**PEL-C**).

Recommended Exposure Limit

The Recommended Exposure Limit [usually abbreviated, **REL**] is still another airborne concentration of some substance of interest which, in this case, the National Institute for Occupational Safety and Health [NIOSH] has researched and developed. Like its two previously listed close counterparts: the **TLV** and the **PEL**, **REL**s are commonly expressed in one of three forms: as an 8-hour Time Weighted Average

(REL-TWA); as a Short Term Exposure Limit (REL-STEL); and as a Ceiling Value (REL-C).

Maximum Concentration Value in the Workplace

The Maximum Concentration Value in the Workplace [usually abbreviated, MAK] is a TLV, PEL, & REL analog, and as such is also an airborne concentration of some substance of interest which, in this case, the Deutsche Forschungsgemeinschaft [DFG], Commission for the Investigation of the Health Hazards of Chemical Compounds in the Work Area, as a branch of the Federal Republic of Germany's central government has developed, adopted, and promulgated. Exactly like its three U. S. counterparts, MAKs are commonly expressed in one of the three standard forms: as an 8-hour Time Weighted Average (MAK-TWA); as a Short Term Exposure Limit (MAK-STEL); and as a Ceiling Value (MAK-C).

Time Weighted Average

The Time Weighted Average [usually abbreviated as a "suffix", -TWA; thus: TLV-TWA, PEL-TWA, REL-TWA, and/or MAK-TWA] is the employee's average airborne exposure in <u>any</u> 8-hour work shift of <u>any</u> 40-hour workweek to which nearly all workers may be repeatedly exposed, day after day, without suffering any adverse effects. It is a value that shall not be exceeded.

Short Term Exposure Limit

The Short Term Exposure Limit [usually abbreviated as a "suffix", -STEL; thus: TLV-STEL, PEL-STEL, REL-STEL, and/or MAK-STEL] is the concentration to which workers can be continuously exposed for short periods of time without suffering from:
1. irritation;
2. chronic or irreversible tissue damage; or
3. narcosis of sufficient degree to increase the likelihood of accidental injury, impair self-rescue, or materially reduce work efficiency – provided also that the corresponding TWA has not been exceeded.

STELs are usually 15-minute (except for those materials for which an alternative time limit has been specified) Time Weighted Average exposures which shall <u>never</u> be exceeded at <u>any</u> time during a work day, even if the corresponding TWA has not been exceeded. In the event any time limit other than 15-minutes is specified for some material or compound, the previous definition still holds, except as modified by the different time limit.

Ceiling Value

The Ceiling Value concentration [usually abbreviated as a "suffix", **-C**; thus: **TLV-C, PEL-C, REL-C**, and/or **MAK-C**] is a concentration that must <u>never</u> be exceeded, even instantaneously, at <u>any</u> time during the work day. In the event that instantaneous monitoring is not feasible, then the Ceiling Value can be assessed as a 15-minute Time Weighted Average exposure which shall not be exceeded at any time during the work day, EXCEPT when the subject vapor can cause immediate irritation with exceedingly short exposures.

Action Level

The Action Level is an 8-hour Time Weighted Average concentration for which there is only a 5% risk of having more than 5% of the employee workdays involve an exposure at a level greater than the relevant **TLV-TWA, PEL-TWA, REL-TEA**, or **MAK-TWA**. This value is most frequently set at or near 50% of the relevant **TLV-TWA, PEL-TWA, REL-TWA**, or **MAK-TWA** concentration standard.

Breathing Zone

The Breathing Zone of an individual is a roughly hemispherical volume immediately forward of that person's shoulders and face, centered roughly on the Adam's Apple, and having a radius of 6 - 9 inches.

Excursion Limit

An Excursion Limit is a term that is frequently called into use in situations, and for substances where no published **STEL** or Ceiling Value exists. It is a Short Term Exposure Limit or Ceiling Value <u>without</u> any legal standing, as would be the case for OSHA published **STEL**s or Ceiling Values. As such, the Excursion Limit is simply an "industry recognized" factor or guideline. For materials that have no published **STEL**, the STEL Excursion Limit is generally understood to be three times the **PEL-TWA**, for no more than 30 minutes during any work day. For materials that have no published Ceiling Value, the Ceiling Value Excursion Limit is generally understood to be five times the **PEL-TWA**, and is treated as a concentration that must <u>never</u> be exceeded at <u>any</u> time.

Upper Explosive Limit & Lower Explosive Limit

The Upper and Lower Explosive Limits (usually abbreviated, **UEL** and **LEL**) refer to the upper and lower vapor concentration boundaries, for some specific compound or material of interest, within which the vapor-air mixture will propagate a flame [ie. explode] if ignited.

Flash Point

The Flash Point of any compound is that temperature to which it must be heated before its vapors can be ignited by a free flame in the presence of air. It is a measure of the flammability of any material, and as such it is a reasonably good criterion for this characteristic. The lower the Flash Point, the more flammable a material is. This value is affected by the relative volatility <u>and</u> the chemical composition of the material in question. Thus, ranked in the order of decreasing flammability, we would find the following to hold true, (for the Flash Point = F): $F_{pure\ hydrocarbons} <$ $F_{oxygenated\ hydrocarbons} < F_{partially\ halogenated\ hydrocarbons}$, etc. For example, the average $F_{gasoline} = -45°$ C, $F_{isopropanol} = 12°$ C, & the average $F_{lubricating\ oil} = 232°$ C, etc. Certain materials (ie. carbon tetrachloride, CCl_4) do not have a Flash Point, since there is <u>no</u> temperature at which their vapors can be ignited.

Immediately Dangerous to Life and/or Health

The Immediately Dangerous to Life and/or Health (usually abbreviated, **IDLH**) Concentration is a concentration level of some substance of interest from which a worker could escape in 30 minutes or less without suffering any escape-impairing symptoms and/or irreversible health effects.

Median Lethal Dose

The Median Lethal Dose (usually abbreviated, **LD50**) is the toxicant dose at which 50% of a population of the same species will die within a specified period of time, under similar experimental conditions. This dosage number is usually expressed as milligrams of toxicant per kilogram of body weight [mg/kg].

Median Effective Dose

The Median Effective Dose (usually abbreviated, **ED50**) is the toxicant dose re-quired to produce a specific non-lethal effect in 50% of a population of the same species, under similar experimental conditions. This dosage number is usually also expressed as milligrams of toxicant per kilogram of body weight [mg/kg].

Median Lethal Concentration

The Median Lethal Concentration (usually abbreviated, **LC50**) is the concentration of toxicant in air which will cause 50% of a population to die within a specified period of time. This factor is a concentration, not a dose, and is usually expressed either as milligrams of toxicant per cubic meter of air [mg/m^3], or as parts per million [ppm].

"Other" Dose-Response, Concentration-Response Terms

Other common analogous Dose-Response or Concentration-Response Terms are as follows – in each case, the definition of each of these terms is directly analogous to the definition of the "Same Name" term listed immediately previously, the only difference being in the percentage figure involved:

Lethal Dose for 10% of a population   (usually abbreviated, **LD10**)

Lethal Dose for 90% of a population   (usually abbreviated, **LD90**)

Effective Dose for 10% of a population   (usually abbreviated, **ED10**)

Effective Dose for 90% of a population   (usually abbreviated, **ED90**)

Lethal Concentration for 10% of a population   (usually abbreviated, **LC10**)

Lethal Concentration for 90% of a population, etc.   (usually abbreviated, **LC90**)

Gas, Vapor, and Aerosol Related

Standard Temperature & Pressure

Standard Temperature and Pressure (usually abbreviated, **STP**) is the designation given to an ambient condition in which, the barometric pressure [**P**] is:

> 1 atmosphere, or
> 760 mm Hg, or
> 14.70 psia, or
> 0.00 psig, or
> 1,013.25 millibars, or
> 760 Torr; and

the ambient temperature [**T**] is:

> 0°C, or
> 32°F, or
> 273.16°K, or
> 459.67°R.

Normal Temperature & Pressure

Normal Temperature and Pressure (usually abbreviated, **NTP**) is the designation given to an ambient condition in which the barometric pressure [**P**] is:

> 1 atmosphere, or
> 760 mm Hg, or
> 14.70 psia, or
> 0.00 psig, or
> 1,013.25 millibars, or
> 760 Torr; and

the ambient temperature [**T**] is:

> 25°C, or
> 77°F, or
> 298.16°K, or
> 536.67°R.

Gas

A Gas is a substance that is in the gaseous state at **NTP**.

Vapor

A Vapor is the gaseous state of any material that would, under **NTP**, be a solid or a liquid.

Aerosol

An Aerosol is a suspension of liquid or solid particles in the air. Particle diameters usually fall in the range: $0.01\mu \leq$ [particle diameter] $\leq 100\mu$.

Terminal or Settling Velocity

The Terminal or Settling Velocity of any suspended particle or aerosol is the average velocity at which that particle falls to earth.

Aerodynamic Diameter

The Aerodynamic Diameter is the diameter of a unit density sphere (ie. density = 1.00 gms/cc) that would have the same settling velocity as the particle or aerosol in question.

Dust

Dust is any particulate material, usually generated by a mechanical process, such as crushing, grinding, etc. Typical dust particles have aerodynamic diameters in the range: $0.5 \ \mu m \leq$ [aerodynamic diameter] $\leq 50 \ \mu m$.

Mist

A Mist is an aerosol suspension of liquid particles in the air, usually formed either by condensation directly from the vapor phase or by some mechanical process. Typical mist droplets have aerodynamic diameters in the range: $40\mu m \leq$ [aerodynamic diameter] $\leq 400 \ \mu m$.

Smoke

Smoke is an aerosol suspension, usually of solid particulates, and usually formed by either the combustion of organic materials or the sublimation of some material. Typical smoke particulates have aerodynamic diameters in the range: $0.01\mu m \leq$ [aerodynamic diameter] $\leq 0.5 \ \mu m$.

Fume

A Fume is an aerosol made up of solid particulates formed by condensation directly from the vapor state. Typical fume particles have aerodynamic diameters in the range: $0.001\,\mu m \leq$ [aerodynamic diameter] $\leq 0.2\,\mu m$.

Aspect Ratio

The Aspect Ratio for any particle is the ratio of its longest or greatest dimension to its shortest or smallest dimension.

Fiber

A Fiber is any particle having an aspect ratio greater than 3.

Ionizing and Non-Ionizing Radiation Related

Electromagnetic Radiation

Electromagnetic Radiation consists of the entire broad spectrum of photonic radiation, from wavelengths of less than 10^{-5} Å to those greater than 10^8 meters. It is comprised of all of the Ionizing and Non-Ionizing Radiation spectral segments, and includes such well known and recognized bands as the following (a partial listing, tabulated from shorter to the longer wavelengths):

> Cosmic Rays
> γ-Rays
> X-Rays
> UV Rays
> Visible Light
> IR Light
> Radar & other types of Microwaves
> Radio.

Ionizing Radiation

Ionizing Radiation is any electromagnetic or particulate radiation – either produced naturally or by some man-made process – that is capable of producing ions, either directly or indirectly, as a result of its [the Ionizing Radiation's] interaction with other forms of matter. Ionizing Radiation can consist of directly or indirectly ionizing particles or a mixture of both. Directly ionizing particles include those that are electrically charged [ie. electrons, positrons, protons, α–particles, etc.], and that have sufficient kinetic energy to produce ionization by collision. In contrast, indirectly ionizing particles are uncharged particles [ie. neutrons, photons, etc.] that can either liberate directly ionizing particles, or initiate some sort of nuclear transition [ie. radioactive decay, fission, etc.] as a result of their interaction with other forms of matter.

Radioactivity

Radioactivity is the process by which certain unstable atomic nuclei undergo nuclear disintegration through the emission of one of the common sub-atomic particles or photons of electromagnetic energy [ie. α–Particles, which are the nuclei of Helium atoms; β–Particles, which are positively or negatively charged electrons; and/or γ–Rays, which are very high energy X-Rays].

Radioactive Decay

Radioactive Decay refers to the actual process – involving one or more separate and distinct steps – by which some specific radioactive element undergoes the transition from its initial condition as an "unstable" nucleus, ultimately to the process termination as a "stable" or non-radioactive nucleus. Radioactive Decay, as implied above occurs in any of several different modes:

Alpha Decay [α–decay],
Beta Decay [β^-–decay],
Positron Decay [β^+–decay],
Gamma Decay [γ–decay],
Neutron Decay [n–decay],
Electron Capture [EC],
Isomeric Transition [IT], and
Internal Conversion [IC].

Measurements of the Strength or Activity of a Source of Radiation

The most common measure of Radiation Source Strength or Activity is the number of radioactive disintegrations that are occurring in the mass of radioactive material per unit time. There are several basic units that are employed in this area; they are listed below, along with their definitions, in terms of disintegrations per second:

Unit of Source Activity	Abbreviation	Disintegrations/min
1 Curie	Ci	2.22×10^{12}
1 Millicurie	mCi	2.22×10^{9}
1 Microcurie	μCi	2.22×10^{6}
1 Picocurie	pCi	2.22
1 Becquerel	Bq	60

Half Life

The Half Life of any radioactive species is the time interval required for the population of that material to be reduced, by Radioactive Decay, to one half of its initial level. In addition to the several classifications listed above, every type of radioactive decay is also characterized by its specific Half Life.

Non-Ionizing Radiation

Non-Ionizing Radiation is electromagnetic spectral radiation having photonic energies of less than 10 - 12 electron volts, which is the approximate threshold energy level required to produce ionization in biological material. Since the energy of any photon of Electromagnetic Radiation is given by Planck's Law as:

$$E = h\nu \qquad \text{where } h = \text{Planck's Constant}$$
$$= 6.63 \times 10^{-27} \text{ erg seconds}$$
$$= 4.14 \times 10^{-15} \text{ electron volt seconds, \&}$$
$$\nu = \text{the frequency of the photon of electromagnetic energy being considered, in sec}^{-1}$$

Therefore, Non-Ionizing Radiation consists of all Electromagnetic Radiation having frequencies of <u>less</u> than 2.42×10^{15} cycles/sec $= 2.42 \times 10^{15}$ sec^{-1}, or wavelengths of <u>greater</u> than 1,240 Å. This includes the <u>entire</u> Electromagnetic Radiation spectrum having wavelengths greater than the lowest energy portions in the UV band (ie. visible light, infrared radiation, microwave radiation, etc.).

Ultraviolet Radiation

Ultraviolet Radiation is that portion of the overall Electromagnetic Radiation spectrum having wavelengths between 10 and 4,000 Å (10^{-9} and 4×10^{-7} meters).

Visible Light + Near, Mid, & Far-Infrared Radiation

Visible Light plus Near, Mid, and Far-Infrared Radiation make up those portion of the overall Electromagnetic Radiation spectrum having wavelengths between 4,000 and 3×10^7 Å (4×10^{-7} and 3×10^{-3} meters). These four specific bands of the overall Electromagnetic Radiation spectrum are made up <u>all</u> the wavelengths that fall in the range between the two wavelengths listed in the following tabulation:

<u>Spectral Band</u>	<u>Shortest Wavelength</u>	<u>Longest Wavelength</u>
Visible Light	4,000 Å	8,000 Å
	4×10^{-7} meters	8×10^{-7} meters
Near-Infrared	8,000 Å	20,000 Å
	8×10^{-7} meters	2×10^{-6} meters
Mid-Infrared	20,000 Å	200,000 Å
	2 microns	20 microns
	2×10^{-6} meters	2×10^{-5} meters
Far-Infrared	200,000 Å	30,000,000 Å
	20 microns	3,000 microns
	2×10^{-5} meters	3×10^{-3} meters

Heat & Cold Stress Related

Wet Bulb Globe Temperature Index

The Wet Bulb Globe Temperature Index [usually abbreviated, **WBGT**] is the most widely used algebraic approximation of an "effective temperature" currently being employed today. It is an Index that can be determined quickly, and with a minimum of effort and operator skill. As an approximation to an "effective temperature", it takes account of virtually all the commonly accepted mechanisms of heat transfer (ie. radiant, evaporative, etc.) <u>except</u> for the cooling effects of wind speed. Because of its simplicity, it has been adopted by the American Conference of Government Hygienists (ACGIH) as its principal index for use in specifying a heat stress related Threshold Limit Value (**TLV**). For outdoor use (ie. in sunshine), the **WGBT** is computed according to the following algebraic sum:

$$\textbf{WBGT} = 0.7\,[\text{NWB}] + 0.2\,[\text{GT}] + 0.1\,[\text{DB}].$$

For indoor use, the WBGT is computed according to the following slightly modified algebraic sum:

$$\textbf{WBGT} = 0.7\,[\text{NWB}] + 0.3\,[\text{GT}].$$

where: [NWB] = Natural Wet-Bulb Temperature,
                [GT]   = Globe Temperature, &
                [DB]   = Dry-Bulb Temperature.

Noise Related

Continuous Noise

An unbroken sound, made up of one or more different frequencies of either constant or varying intensity or sound level, is referred to as Continuous Noise. If a Continuous Noise is constant and unvarying in its amplitude, it would be referred to as "steady" Continuous Noise. Its alternative would be "varying" Continuous Noise. Continuous Noise is a fairly commonly occurring phenomenon – both in the industrial and the natural environment.

In the natural environment, one might regard the sound of a waterfall as "steady" Continuous Noise, while the sounds of wind blowing through a forest would be "varying" Continuous Noise.

In the industrial environment, the sound of a rotating electric motor (ie. a fan, a pump, etc.) would be "steady" Continuous Noise, while the operation of a floor waxer, relative to a fixed observer, would be "varying" Continuous Noise.

Continuous Noise is an extremely useful concept. In assessing the potential hazards of any noise filled environment, one attempts to quantify the existing noise pattern in terms of the "steady" Continuous Noise that could, in theory, replace it without altering any of the adverse effects that might be being experienced by a human observer. For any environment, an $L_{equivalent}$, or the "steady" Continuous Noise level, described above, can usually be determined; the sound intensity level of this "steady" Continuous Noise – which equals $L_{equivalent}$ – can then be used to evaluate the overall sound hazard that is posed to individuals who must occupy that environment.

Intermittent Noise

Intermittent Noise is a broken or non-continuous sound (ie. sound bursts or periods of time during which there are intervals of quiet [non-sound] and subsequent intervals during which there is measurable sound). Intermittent Noise can also be made up of one or more different frequencies of sound, of either constant or varying intensity or sound level. The sound of an operating typewriter would be considered an intermittent noise.

Sound Frequency

The Frequency of any Sound is the time rate at which complete cycles of high and low pressure regions are produced by the source of the sound. The most common unit of Sound Frequency is the number of cycles/second or the Hertz (abbreviated,

Hz). The frequency range of the human ear varies with age and circumstance; but a normal hearing "young" ear will usually be able to hear sounds, at moderate levels, in the range, 20 - 20,000 Hz = 20 - 20,000 cycles/second.

Sound Wavelength

The Wavelength of a Sound is the precise distance required for one complete pressure cycle (ie. one cycle of high and low pressure regions) for that frequency of sound.

Pitch

The Pitch of a sound is the subjective auditory perception of the frequency of that sound. It, of course, depends upon the frequency of the sound, but also on its waveform and overall sound pressure level.

Velocity of Sound

The Velocity of Sound is the speed at which the regions of high and low pressure move away from the source of the sound. For all practical purposes this Velocity can be considered to be a constant in whatever medium the sound is transiting. It varies directly as the square root of the density of the medium involved, and inversely as the compressibility of that medium. For example:

Medium	Velocity of Sound
air	~ 1,130 ft/sec
sea water	~ 4,680 ft/sec
hard wood	~ 13,040 ft/sec
steel	~ 16,550 ft/sec

Loudness

The Loudness of a Sound is an observer's impression of its amplitude. This subjective judgment is influenced strongly by the characteristics of the ear that is doing the hearing.

Sound Intensity

The Sound Intensity of any sound source at any particular location is the average rate at which sound energy from that source is being transmitted through a unit area that is normal to the direction in which the sound is propagating. The most common units of measure for Sound Intensity are joules/m^2/sec, which also equals watts/m^2.

Sound Intensity is usually expressed as a Sound Intensity Level, in decibels referenced to 10^{-12} watts/m^2.

Sound Power

The Sound Power of any sound source is the total sound energy radiated by that source per unit time. The most common units of measure for Sound Power are watts. Sound Power is usually expressed as a Sound Power Level, in decibels referenced to 10^{-12} watts.

Sound Pressure

Sound Pressure normally refers to the RMS values of the pressure changes, above and below atmospheric pressure, which are used to measure steady state or continuous noise. The most common units of measure for Sound Pressure are:

$$\text{newtons per square meter} = n/m^2$$
$$\text{dynes per square centimeter} = d/cm^2$$
$$\text{microbars}$$

Sound Pressure is usually expressed as a Sound Pressure Level, in decibels referenced to 2×10^{-5} n/m^2.

Root-Mean-Square [RMS] Sound Pressure

The RMS value of any changing quantity, such as sound pressure, is the square root of the mean of the squares of instantaneous values of that quantity.

Sound Measurement Time Weightings

In the quantification of various Sound Pressure Levels [in decibels], there are four different commonly used averaging periods, or time weightings, that are commonly used, as part of the standard RMS detection method. These four are: Peak, Impulse, Fast, and Slow Noise Weightings.

Peak Noise

Sound having a duration of less than 100 milliseconds is considered to be in the Peak Noise category. Such sounds will also fall under the category of Impulsive or Impact Noise.

Impulsive or Impact Noise

The type of noise produced by such things as a gun being fired, the operation of an industrial punch press, or the use of a hammer to drive a nail are all highly transient sound phenomena, and are usually treated as Impulsive or Impact Noises. The definition of this type of noise is any sound that has an amplitude rise time of 35 milliseconds or less, and a fall time of 1,500 milliseconds or less is Impulsive or Impact Noise.

Fast Time Weighted Noise

Sound pressure level measurements using a 125 millisecond moving average weighting period are said to have been determined using Fast Time Weighting.

Slow Time Weighted Noise

Sound pressure level measurements using a 1.0 second moving average weighting period are said to have been determined using Slow Time Weighting.

Ventilation Related

Static Pressure

The Static Pressure is the pressure exerted in all directions by a fluid at rest. For any fluid in motion, this parameter is measured in a direction <u>normal</u> to the direction of fluid flow. Static Pressure is usually expressed in "inches of water gauge" when dealing with air flowing in a duct; this parameter can be either negative or positive.

Velocity Pressure

The Velocity Pressure is the kinetic pressure, in the direction of fluid flow, that is necessary to cause that fluid to flow at a particular velocity. Velocity Pressure is usually expressed in "inches of water gauge" when dealing with air flowing in a duct; this parameter is always positive. It is measured in a direction <u>parallel</u> to the direction of fluid flow, looking "upstream".

Total Pressure

The Total Pressure is the algebraic sum of the Static Pressure and the Velocity Pressure. It, too, is usually expressed in "inches of water gauge" when dealing with air flowing in a duct.

Vapor Pressure

The Vapor Pressure is the static pressure exerted by a vapor. If the vapor phase of some material of interest is in equilibrium with the liquid phase of that same material, then the Vapor Pressure that exists can be and is usually referred to as the "Saturated Vapor Pressure" for that material. The Saturated Vapor Pressure for any material is solely dependent on the temperature of that material. Frequently "Vapor Pressure" and "Saturated Vapor Pressure" are used synonymously.

Entry Loss

The Entry Loss is the pressure loss incurred by a fluid as it flows into a duct or hood opening. Entry Losses are usually expressed in "inches of water gauge".

Absolute Humidity

The Absolute Humidity is the weight of the vaporous water in the air per unit volume of air, usually expressed in lbs/ft^3 or $grams/cm^3$.

Relative Humidity

Relative Humidity is the ratio of the actual or measured Vapor Pressure of water in an air mass, to the Saturated Vapor Pressure of pure water at the same temperature as that of the air mass. Relative Humidity can also be thought of as the percentage that the actual measured absolute concentration of water in the air under <u>any</u> ambient conditions is to the saturation concentration of water in air under <u>the</u> <u>same</u> ambient conditions.

Manometer

A Manometer is an instrument for measuring the pressure in a fluid or gas. A Manometer is usually a U-tube filled with water, light oil, or mercury. This implement is usually constructed so that the observed displacement of liquid in it will indicate the pressure being exerted on it by the fluid being monitored. In a Static Pressure measuring situation, the <u>plane</u> of the opening of the Manometer tube in the fluid being monitored will be <u>parallel</u> to the direction of flow of that fluid (ie. the Manometer tube, itself, will be situated <u>perpendicular</u> to the flow of that fluid). In a Velocity Pressure measuring situation, the <u>plane</u> of the opening of the Manometer tube in the fluid being monitored will be <u>perpendicular</u> to the direction of flow of that fluid, and will be directed so that the opening at the end of the tube faces directly into that flow (ie. the Manometer tube, itself, will be situated <u>parallel</u> to the flow of that fluid, with the tube opening facing "upstream").

Air Horsepower

The Air Horsepower is the theoretical horsepower required to operate a fan, assuming there are no losses in the fan (ie. the horsepower required to operate a 100% efficient fan).

Brake Horsepower

The Brake Horsepower is the horsepower actually required to operate a fan. Brake Horsepower involves <u>all</u> the internal losses in the fan and can only be measured by conducting an actual test on the fan.

Hood

A Hood is a shaped inlet designed to capture contaminated air and conduct it into the exhaust duct system. Hoods can be plain or flanged.

Standard Air

Standard Air, as Ventilation Engineers view it, differs from air at STP or Standard Conditions (as defined earlier). Standard Air is dry air at a temperature of:

> 70°F, or
> 21.1°C; and

a barometric pressure of:

> 29.92 in. of Hg, or
> 1.00 atmosphere.

Standard air weighs 0.075 lbs/ft^3, and has a specific heat of 0.24 BTU/lb/°F.

Capture Velocity

The Capture Velocity is the air velocity at any point in front of a hood, or hood opening, that is necessary to overcome the opposing air currents and other factors, and thereby "capture" the contaminated air at that point by causing it to flow or be drawn into the hood.

Coefficient of Entry

The Coefficient of Entry is the actual rate of flow produced by a given Hood Static Pressure, when compared to the theoretical flow that would result if the Static Pressure in the Hood could be converted, with 100% efficiency, into the Velocity Pressure of the fluid being drawn into the Hood. The Coefficient of Entry is also the ratio of the actual flow rate to the theoretical maximum flow rate.

Blast Gate

A Blast Gate is a sliding duct damper whose purpose is to permit easy adjustment of the flow volume in that duct.

Statistics & Probability Related

Population

The Population is the entire set or family of objects, data, measurements, etc. being considered from a statistical, probabilistic, or combinatorial perspective.

Variable

A Variable is a characteristic or property of an individual member of a population. The name, variable, is derived from the fact that any particular characteristic of interest may "vary" among the individual members of a population.

Sample

A Sample is a subset of the members of an entire population.

Frequency Distribution

A Frequency Distribution is made up of a tabulation of any of the variable characteristics of any population that can be measured, counted, tabulated, or correlated.

Mean

The Mean of any set of variable data is the sum of the individual variables of the items of that data set, divided by the total number of items in the set. It is the average value for the set of data, and is the first important measure of the central tendency of that set of variables.

Median

The Median of any set of variable data is the middle value of that data set, when all the individual variable members of the set have been arranged either in ascending or descending order. It is the second important measure of the central tendency of that set of variables.

Mode

The Mode of any set of variable data points is the value of the most common or frequently occurring member of that set.

Range

The Range of any set of variable data is the value of the largest member of that set minus the value of the smallest member of that same set.

Sample Variance

The Sample Variance of any set of variable data points containing a total of **n** values is equal to the sum of the squared distances of each member of that set from the set's mean value, divided by **(n - 1)**.

Sample Standard Deviation

The Sample Standard Deviation of any set of variable data points is equal to the positive square root of the sample variance, as defined above.

Population Variance

The Population Variance of any set of variable data points containing a total of **n** values is equal to the <u>average</u> of the squared distances of each member of that set from the set's mean value — this parameter differs from the Sample Variance defined above by virtue of the fact that the sum of the squared distances of each member of this set from the set's mean value is divided by **n**, rather than **(n - 1)**.

Population Standard Deviation

The Population Standard Deviation of any set of variable data points is equal to the positive square root of the population variance, as defined above.

Normal Distribution

A Normal Distribution is the frequency distribution that occurs most commonly in nature and industry — ie. the heights of adult males in California, the average daily rainfall in the Amazon River basin over a decade, the accident rate among interstate truck drivers.

Conversion Factors

Physical Constants — Alphabetic Listing

1 atm (atmosphere) =
    1.013 bars
    10.133 newtons/cm^2 (newtons/square centimeter)
    33.90 ft. of H_2O (feet of water)
    101.325 kp (kilopascals)
    1,013.25 mb (millibars)
    14.70 psia (pounds/square inch - absolute)
    760 torr
    760 mm Hg (millimeters of mercury)

1 bar =
    0.987 atm (atmospheres)
    1×10^6 dynes/cm^2 (dynes/square centimeter)
    33.45 ft. of H_2O (feet of water)
    1×10^5 pascals [nt/m^2] (newtons/square meter)
    750.06 torr
    750.06 mm Hg (millimeters of mercury)

1 Bq (becquerel) =
    1 radioactive disintegration/second
    2.7×10^{-11} Ci (curie)
    2.7×10^{-8} mCi (millicurie)

1 BTU (British Thermal Unit) =
    252 cal (calories)
    1,055.06 j (joules)
    10.41 liter-atmospheres
    0.293 watt-hours

1 cal (calorie) =
    3.97×10^{-3} BTUs (British Thermal Units)
    4.18 j (joules)
    0.0413 liter-atmospheres
    1.163×10^{-3} watt-hours

1 cm (centimeter) =

0.0328 ft (feet)
0.394 in (inches)
10,000 microns (micrometers)
100,000,000 Å = 10^8 Å (Ångstroms)

1 cc (cubic centimeter) =

3.53×10^{-5} ft^3 (cubic feet)
0.061 in^3 (cubic inches)
2.64×10^{-4} gal (gallons)
0.001 l (liters)
1.00 ml (milliliters)

1 ft^3 (cubic foot) =

28,317 cc (cubic centimeters)
1,728 in^3 (cubic inches)
0.0283 m^3 (cubic meters)
7.48 gal (gallons)
28.32 l (liters)
29.92 qts (quarts)

1 in^3 (cubic inch) =

16.39 cc (cubic centimeters)
16.39 ml (milliliters)
5.79×10^{-4} ft^3 (cubic feet)
1.64×10^{-5} m^3 (cubic meters)
4.33×10^{-3} gal (gallons)
0.0164 l (liters)
0.55 fl oz (fluid ounces)

1 m^3 (cubic meter) =

1,000,000 cc = 10^6 cc (cubic centimeters)
35.31 ft^3 (cubic feet)
61,023 in^3 (cubic inches)
264.17 gal (gallons)
1,000 l (liters)

1 yd^3 (cubic yard) =

201.97 gal (gallons)
764.55 l (liters)

1 Ci (curie) =

3.7×10^{10} radioactive disintegrations/second
3.7×10^{10} Bq (becquerel)
1,000 mCi (millicurie)

1 day =

24 hrs (hours)
1,440 min (minutes)
86,400 sec (seconds)
0.143 weeks
$2,738 \times 10^{-3}$ yrs (years)

1°C (expressed as an interval) =

$1.8°F = \left[\dfrac{9}{5}\right]°F$ (degrees Celsius)

1.8°R (degrees Rankine)
1.0°K (degrees Kelvin)

°C (degree Celsius) =

$\left[\dfrac{5}{9}\right] \times [°F - 32°]$ (degrees Celsius)

1°F (expressed as an interval) =

$0.556°C = \left[\dfrac{5}{9}\right]°C$ (degrees Fahrenheit)

1.0°R (degrees Rankine)
0.556°K (degrees Kelvin)

°F (degree Fahrenheit) =

$\left[\dfrac{9}{5}\right] \times [°C] + 32°$ (degrees Fahrenheit)

1 dyne =

1×10^{-5} nt (newtons)

1 ev (electron volt) =

1.602×10^{-12} ergs
1.602×10^{-19} j (joules)

1 erg =

1 dyne-centimeters
1×10^{-7} j (joules)
2.78×10^{-11} watt-hours

1 ft (foot) =	30.48 cm (centimeters)
	12 in (inches)
	0.3048 m (meters)
	1.65×10^{-4} nt (nautical miles)
	1.89×10^{-4} mi (statute miles)
1 fps (feet/second) =	1.097 kmph (kilometers/hour)
	0.305 mps (meters/second)
	0.01136 mph (miles/hour)
1 gal (gallon) =	3,785 cc (cubic centimeters)
	0.134 ft^3 (cubic feet)
	231 in^3 (cubic inches)
	3.785 l (liters)
1 gm (gram) =	0.001 kg (kilograms)
	1,000 mg (milligrams)
	1,000,000 ng = 10^6 ng (nanograms)
	2.205×10^{-3} lbs (pounds)
1 gm/cc (grams/cubic centimeter) =	62.43 lbs/ft^3 (pounds/cubic foot)
	0.0361 lbs/in^3 (pounds/cubic inch)
	8.345 lbs/gal (pounds/gallon)
1 Gy (gray) =	1 j/kg (joules/kilogram)
	100 rad
	1 Sv (sievert) [unless modified through division by various factors , such as Q & N]
1 hp (horsepower) =	745.7 j/sec (joules/sec)
1 hr (hour) =	0.0417 days
	60 min (minutes)
	3,600 sec (seconds)
	5.95×10^{-3} weeks
	1.14×10^{-4} yrs (years)

1 in (inch) =	2.54 cm (centimeters)
	1,000 mils
1 inch of water =	1.86 mm Hg (millimeters of mercury)
	249.09 pascals
	0.0361 psi (lbs/in^2)
1 j (joule) =	9.48x10^{-4} BTUs (British Thermal Units)
	0.239 cal (calories)
	10,000,000 ergs = 1x10^7 ergs
	9.87x10^{-3} liter-atmospheres
	1.00 nt-m (newton-meters)
1 kcal (kilocalorie) =	3.97 BTUs (British Thermal Units)
	1,000 cal (calories)
	4,186.8 j (joules)
1 kg (kilogram) =	1,000 gms (grams)
	2.205 lbs (pounds)
1 km (kilometer) =	3,280.8 ft (feet)
	0.54 nt (nautical miles)
	0.6214 mi (statute miles)
1 kw (kilowatt) =	56.87 BTU/min (British Thermal Units/minute)
	1.341 hp (horsepower)
	1,000 j/sec (joules/sec)
1 kw-hr (kilowatt-hour) =	3,412.14 BTU (British Thermal Units)
	3.6x10^6 j (joules)
	859.8 kcal (kilocalories)

1 l (liter) =

1,000 cc (cubic centimeters)
1 dm^3 (cubic decimeters)
0.0353 ft^3 (cubic feet)
61.02 in^3 (cubic inches)
0.264 gal (gallons)
1,000 ml (milliliters)
1.057 qts (quarts)

1 m (meter) =

1×10^{10} Å (Ångstroms)
100 cm (centimeters)
3.28 ft (feet)
39.37 in (inches)
1×10^{-3} km (kilometers)
1,000 mm (millimeters)
$1,000,000 \; \mu = 1 \times 10^6 \; \mu$ (micrometers)
1×10^9 nm (nanometers)

1 mps (meters/second) =

196.9 fpm (feet/minute)
3.6 kmph (kilometers/hour)
2.237 mph (miles/hour)

1 kt (nautical mile) =

6,076.1 ft (feet)
1.852 km (kilometers)
1.15 mi (statute miles)
2,025.4 yds (yards)

1 mi (statute mile) =

5,280 ft (feet)
1.609 km (kilometers)
1,609.3 m (meters)
0.869 nt (nautical miles)
1,760 yds (yards)

1 mph (mile/hour) =

88 fpm (feet/minute)
1.61 kmph (kilometers/hour)
0.447 mps (meters/second)

1 mCi (millicurie) =

0.001 Ci (curie)
3.7×10^{10} radioactive disintegrations/second
3.7×10^{10} Bq (becquerel)

1 mm Hg (millimeter of mercury) =

1.316×10^{-3} atm (atmospheres)
0.535 in H_2O (inches of water)
1.33 mb (millibars)
133.32 pascals
1 torr
0.0193 psia (pounds/square inch - absolute)

1 min (minute) =

6.94×10^{-4} days
0.0167 hrs (hours)
60 sec (seconds)
9.92×10^{-5} weeks
1.90×10^{-6} yrs (years)

1 nt (newton) =

1×10^5 dynes

1 nt-m (newton-meter) =

1.00 j (joules)
2.78×10^{-4} watt-hours

1 ppm[wt] (parts/million-weight) =

1.00 mg/kg (milligrams/kilogram)

1 ppm (parts/million-volume) =

1.00 ml/m^3 (milliliters/cubic meter)

1 pascal =

9.87×10^{-6} atm (atmospheres)
4.015×10^{-3} in H_2O (inches of water)
0.01 mb (millibars)
7.5×10^{-3} mm Hg (millimeters of mercury)

1 lbs (pound) =

453.59 gms (grams)
16 oz (ounces)

1 lbs/ft^3 (pounds/cubic foot) =

16.02 gms/l (grams/liter)

1 lbs/in^3 (pounds/cubic inch) =

27.68 gms/cc (grams/cubic centimeter)
1,728 lbs/ft^3 (pounds/cubic foot)

1 psi (pounds/square inch) =

0.068 atm (atmospheres)
27.67 in H_2O (inches of water)
68.85 mb (millibars)
51.71 mm Hg (millimeters of mercury)
6,894.76 pascals

1 qt (quart) =

946.4 cc (cubic centimeters)
57.75 in^3 (cubic inches)
0.946 1 (liters)

1 rad =

100 ergs/gm (ergs/gram)
0.01 Gy (gray)
1 rem [unless modified through division by
  various factors , such as Q & N]

1 rem =

1 rad [unless modified through multiplication
  by various factors, such as Q & N]

1 Sv (sievert) =

1 Gy (gray) [unless modified through division by various factors , such as Q & N]

1 cm^2 (square centimeter) =

1.076×10^{-3} ft^2 (square feet)
0.155 in^2 (square inches)
1×10^{-4} m^2 (square meters)

1 ft^2 (square foot) =

2.296×10^{-5} acres
929.03 cm^2 (square centimeters)
144 in^2 (square inches)
0.0929 m^2 (square meters)

1 m^2 (square meter) =

10.76 ft^2 (square feet)
1,550 in^2 (square inches)

1 mi^2 (square mile) =

640 acres
2.79×10^7 ft^2 (square feet)
2.59×10^6 m^2 (square meters)

1 torr =   1.33 mb (millibars)

1 watt =   3.41 BTU/hr (British Thermal Units/hour)
                                1.341×10^{-3} hp (horsepower)
                                1.00 j/sec (joules/second)

1 watt-hour =   3.412 BTUs (British Thermal Units)
                                859.8 cal (calories)
                                3,600 j (joules)
                                35.53 liter-atmospheres

1 week =   7 days
                                168 hrs (hours)
                                10,080 min (minutes)
                                $6,048 \times 10^{5}$ sec (seconds)
                                0.0192 yrs (years)

1 yr (year) =   365.25 days
                                8,766 hrs (hours)
                                5.26×10^{5} min (minutes)
                                3.16×10^{7} sec (seconds)
                                52.18 weeks

Physical Constants — Listed by Type of Unit

Units of Length

1 cm (centimeter) =

0.0328 ft (feet)
0.394 in (inches)
10,000 microns (micrometers)
100,000,000 Å = 10^8 Å (Ångstroms)

1 ft (foot) =

30.48 cm (centimeters)
12 in (inches)
0.3048 m (meters)
1.65×10^{-4} nt (nautical miles)
1.89×10^{-4} mi (statute miles)

1 in (inch) =

2.54 cm (centimeters)
1,000 mils

1 km (kilometer) =

3,280.8 ft (feet)
0.54 nt (nautical miles)
0.6214 mi (statute miles)

1 m (meter) =

1×10^{10} Å (Ångstroms)
100 cm (centimeters)
3.28 ft (feet)
39.37 in (inches)
1×10^{-3} km (kilometers)
1,000 mm (millimeters)
$1,000,000 \, \mu = 1 \times 10^6 \, \mu$ (micrometers)
1×10^9 nm (nanometers)

1 kt (nautical mile) =

6,076.1 ft (feet)
1.852 km (kilometers)
1.15 mi (statute miles)
2,025.4 yds (yards)

1 mi (statute mile) =

5,280 ft (feet)
1.609 km (kilometers)
1,609.3 m (meters)
0.869 nt (nautical miles)
1,760 yds (yards)

Units of Area

1 cm^2 (square centimeter) =

1.076×10^{-3} ft^2 (square feet)
0.155 in^2 (square inches)
1×10^{-4} m^2 (square meters)

1 ft^2 (square foot) =

2.296×10^{-5} acres
929.03 cm^2 (square centimeters)
144 in^2 (square inches)
0.0929 m^2 (square meters)

1 m^2 (square meter) =

10.76 ft^2 (square feet)
1,550 in^2 (square inches)

1 mi^2 (square mile) =

640 acres
2.79×10^7 ft^2 (square feet)
2.59×10^6 m^2 (square meters)

Units of Volume

1 cc (cubic centimeter) =

3.53×10^{-5} ft^3 (cubic feet)
0.061 in^3 (cubic inches)
2.64×10^{-4} gal (gallons)
0.001 l (liters)
1.00 ml (milliliters)

1 ft^3 (cubic foot) =	28,317 cc (cubic centimeters)
	1,728 in^3 (cubic inches)
	0.0283 m^3 (cubic meters)
	7.48 gal (gallons)
	28.32 l (liters)
	29.92 qts (quarts)
1 in^3 (cubic inch) =	16.39 cc (cubic centimeters)
	16.39 ml (milliliters)
	5.79x10^{-4} ft^3 (cubic feet)
	1.64x10^{-5} m^3 (cubic meters)
	4.33x10^{-3} gal (gallons)
	0.0164 l (liters)
	0.55 fl oz (fluid ounces)
1 m^3 (cubic meter) =	1,000,000 cc = 10^6 cc (cubic centimeters)
	35.31 ft^3 (cubic feet)
	61,023 in^3 (cubic inches)
	264.17 gal (gallons)
	1,000 l (liters)
1 yd^3 (cubic yard) =	201.97 gal (gallons)
	764.55 l (liters)
1 gal (gallon) =	3,785 cc (cubic centimeters)
	0.134 ft^3 (cubic feet)
	231 in^3 (cubic inches)
	3.785 l (liters)
1 l (liter) =	1,000 cc (cubic centimeters)
	1 dm^3 (cubic decimeters)
	0.0353 ft^3 (cubic feet)
	61.02 in^3 (cubic inches)
	0.264 gal (gallons)
	1,000 ml (milliliters)
	1.057 qts (quarts)

1 qt (quart) =   946.4 cc (cubic centimeters)
                                   57.75 in^3 (cubic inches)
                                   0.946 l (liters)

Units of Mass

1 gm (gram) =   0.001 kg (kilograms)
                                   1,000 mg (milligrams)
                                   1,000,000 ng = 10^6 ng (nanograms)
                                   2.205x10^{-3} lbs (pounds)

1 kg (kilogram) =   1,000 gms (grams)
                                   2.205 lbs (pounds)

1 lbs (pound) =   453.59 gms (grams)
                                   16 oz (ounces)

Units of Time

1 day =   24 hrs (hours)
                                   1,440 min (minutes)
                                   86,400 sec (seconds)
                                   0.143 weeks
                                   2,738x10^{-3} yrs (years)

1 hr (hour) =   0.0417 days
                                   60 min (minutes)
                                   3,600 sec (seconds)
                                   5.95x10^{-3} weeks
                                   1.14x10^{-4} yrs (years)

1 min (minute) =   6.94x10^{-4} days
                                   0.0167 hrs (hours)
                                   60 sec (seconds)
                                   9.92x10^{-5} weeks
                                   1.90x10^{-6} yrs (years)

1 week =	7 days
	168 hrs (hours)
	10,080 min (minutes)
	$6,048 \times 10^5$ sec (seconds)
	0.0192 yrs (years)
1 yr (year) =	365.25 days
	8,766 hrs (hours)
	5.26×10^5 min (minutes)
	3.16×10^7 sec (seconds)
	52.18 weeks

Units of the Measure of Temperature

1°C (expressed as an interval) = $\quad 1.8°F = \left[\dfrac{9}{5}\right]°F$ (degrees Celsius)

$\qquad\qquad\qquad\qquad\qquad$ 1.8°R (degrees Rankine)

$\qquad\qquad\qquad\qquad\qquad$ 1.0°K (degrees Kelvin)

°C (degree Celsius) = $\qquad \left[\dfrac{5}{9}\right] \times [°F - 32°]$ (degrees Celsius)

1°F (expressed as an interval) = $\quad 0.556°C = \left[\dfrac{5}{9}\right]°C$ (degrees Fahrenheit)

$\qquad\qquad\qquad\qquad\qquad$ 1.0°R (degrees Rankine)

$\qquad\qquad\qquad\qquad\qquad$ 0.556°K (degrees Kelvin)

°F (degree Fahrenheit) = $\qquad \left[\dfrac{9}{5}\right] \times [°C] + 32°$ (degrees Fahrenheit)

Units of Force

1 dyne =	1×10^{-5} nt (newtons)
1 nt (newton) =	1×10^5 dynes

Units of Work or Energy

1 BTU (British Thermal Unit) =
    252 cal (calories)
    1,055.06 j (joules)
    10.41 liter-atmospheres
    0.293 watt-hours

1 cal (calorie) =
    3.97×10^{-3} BTUs (British Thermal Units)
    4.18 j (joules)
    0.0413 liter-atmospheres
    1.163×10^{-3} watt-hours

1 ev (electron volt) =
    1.602×10^{-12} ergs
    1.602×10^{-19} j (joules)

1 erg =
    1 dyne-centimeters
    1×10^{-7} j (joules)
    2.78×10^{-11} watt-hours

1 j (joule) =
    9.48×10^{-4} BTUs (British Thermal Units)
    0.239 cal (calories)
    10,000,000 ergs = 1×10^{7} ergs
    9.87×10^{-3} liter-atmospheres
    1.00 nt-m (newton-meters)

1 kcal (kilocalorie) =
    3.97 BTUs (British Thermal Units)
    1,000 cal (calories)
    4,186.8 j (joules)

1 kw-hr (kilowatt-hour) =
    3,412.14 BTU (British Thermal Units)
    3.6×10^{6} j (joules)
    859.8 kcal (kilocalories)

1 nt-m (newton-meter) =
    1.00 j (joules)
    2.78×10^{-4} watt-hours

1 watt-hour =

3.412 BTUs (British Thermal Units)
859.8 cal (calories)
3,600 j (joules)
35.53 liter-atmospheres

Units of Power

1 hp (horsepower) =

745.7 j/sec (joules/sec)

1 kw (kilowatt) =

56.87 BTU/min (British Thermal
Units/minute)
1.341 hp (horsepower)
1,000 j/sec (joules/sec)

1 watt =

3.41 BTU/hr (British Thermal Units/hour)
1.341×10^{-3} hp (horsepower)
1.00 j/sec (joules/second)

Units of Pressure

1 atm (atmosphere) =

1.013 bars
10.133 newtons/cm^2 (newtons/square centimeter)
33.90 ft. of H$_2$O (feet of water)
101.325 kp (kilopascals)
1,013.25 mb (millibars)
14.70 psia (pounds/square inch - absolute)
760 torr
760 mm Hg (millimeters of mercury)

1 bar =

0.987 atm (atmospheres)
1×10^6 dynes/cm^2 (dynes/square centimeter)
33.45 ft. of H$_2$O (feet of water)
1×10^5 pascals [nt/m^2] (newtons/square
meter)
750.06 torr
750.06 mm Hg (millimeters of mercury)

1 inch of water =	1.86 mm Hg (millimeters of mercury)
	249.09 pascals
	0.0361 psi (lbs/in^2)
1 mm Hg (millimeter of mercury) =	1.316×10^{-3} atm (atmospheres)
	0.535 in H_2O (inches of water)
	1.33 mb (millibars)
	133.32 pascals
	1 torr
	0.0193 psia (pounds/square inch - absolute)
1 pascal =	9.87×10^{-6} atm (atmospheres)
	4.015×10^{-3} in H_2O (inches of water)
	0.01 mb (millibars)
	7.5×10^{-3} mm Hg (millimeters of mercury)
1 psi (pounds/square inch) =	0.068 atm (atmospheres)
	27.67 in H_2O (inches of water)
	68.85 mb (millibars)
	51.71 mm Hg (millimeters of mercury)
	6,894.76 pascals
1 torr =	1.33 mb (millibars)

Units of Velocity or Speed

1 fps (feet/second) =	1.097 kmph (kilometers/hour)
	0.305 mps (meters/second)
	0.01136 mph (miles/hour)
1 mps (meters/second) =	196.9 fpm (feet/minute)
	3.6 kmph (kilometers/hour)
	2.237 mph (miles/hour)
1 mph (mile/hour) =	88 fpm (feet/minute)
	1.61 kmph (kilometers/hour)
	0.447 mps (meters/second)

Units of Density

1 gm/cc (grams/cubic centimeter) =   62.43 lbs/ft^3 (pounds/cubic foot)
                                     0.0361 lbs/in^3 (pounds/cubic inch)
                                     8.345 lbs/gal (pounds/gallon)

1 lbs/ft^3 (pounds/cubic foot) =   16.02 gms/l (grams/liter)

1 lbs/in^3 (pounds/cubic inch) =   27.68 gms/cc (grams/cubic centimeter)
                                     1,728 lbs/ft^3 (pounds/cubic foot)

Units of Concentration

1 ppm[wt] (parts/million-weight) =   1.00 mg/kg (milligrams/kilogram)

1 ppm (parts/million-volume) =   1.00 ml/m^3 (milliliters/cubic meter)

Radiation & Dose Related Units

1 Bq (becquerel) =   1 radioactive disintegration/second
                                     2.7×10^{-11} Ci (curie)
                                     2.7×10^{-8} mCi (millicurie)

1 Ci (curie) = 3.7×10^{10} radioactive disintegrations/second
                                     3.7×10^{10} Bq (becquerel)
                                     1,000 mCi (millicurie)

1 Gy (gray) =   1 j/kg (joule/kilogram)
                                     100 rad
                                     1 Sv (sievert) [unless modified through divi-
                                         sion by various factors , such as Q & N]

1 mCi (millicurie) =   0.001 Ci (curie)
                                     3.7×10^{10} radioactive disintegrations/second
                                     3.7×10^{10} Bq (becquerel)

1 rad =   100 ergs/gm (ergs/gm)
                        0.01 Gy (gray)
                        1 rem [unless modified through division by
                          various factors , such as Q & N]

1 rem =   1 rad [unless modified through multiplication
                          by various factors, such as Q & N]

1 Sv (sievert) =   1 Gy (gray) [unless modified through divi-
                          sion by various factors , such as Q & N]

Metric Prefixes (for use with *SI* Units)

"EXPANDING" PREFIXES   "DIMINISHING" PREFIXES

Prefix	Abbreviation	Multiplier	Prefix	Abbreviation	Multiplier
deca-	da	10^1	deci-	d	10^{-1}
hecto-	h	10^2	centi-	c	10^{-2}
kilo-	k	10^3	milli-	m	10^{-3}
mega-	M	10^6	micro-	μ	10^{-6}
giga-	G	10^9	nano-	n	10^{-9}
tera-	T	10^{12}	pico-	p	10^{-12}
peta-	P	10^{15}	femto-	f	10^{-15}
exa-	E	10^{18}	atto-	a	10^{-18}

Formulae

Worker Safety Related:

Calculations Involving Time Weighted Averages [TWAs]:

Equation 1A:

The following Equation, #1A, references a **Time Weighted Average** calculation: in this specific case, for determining an **Eight-Hour Threshold Limit Value - Time Weighted Average** [ie. the **TLV-TWA**]. The identical Equation applies to any Time Weighted Average determination; it can be utilized for any total **Base Time Period** or **Time Interval**, and for any of the currently published parameters (e.g. for the determination of an 8-hour, or a 15-minute, or any other **Time Interval** based Time Weighted Average, for any of the established parameters: the **TLV**, the **PEL**, the **REL**, or the **MAK**). As an example, a Short Term Exposure Limit determination [ie. the **PEL-STEL**] uses the identical Equation — namely, **#1A** — with only the denominator differing from that in the following example Formula, for which the **Time Interval** is 8-hours. The denominator would be determined by the specified **STEL Time Interval** for whatever material is involved in the determination (ie. if the material being evaluated was isopropanol, then the **PEL-STEL Time Interval** would be 15 minutes):

$$TLV - TWA = \frac{\sum_{i=1}^{n} T_i C_i}{\sum_{i=1}^{n} T_i} = \frac{T_1 C_1 + T_2 C_2 + \ldots + T_n C_n}{T_1 + T_2 + \ldots + T_n}$$

Where: $TLV - TWA$ = the **Threshold Limit Value - Time Weighted Average Concentration** that existed during the **Time Intervals**, the T_i's, used in the evaluation,

T_i = the ith **Time Interval** in the overall time period, (& for the **TLV-TWA**, the $\sum_{i=1}^{n} T_i$ = 8 hours), &

C_i = the ith **Concentration Value** (of the single component of interest) that existed during the ith **Time Interval**, T_i.

Equation **1B**:

The following Equation, **#1B**, is used to determine the **Effective Percent Exposure Level** for <u>any</u> of the established parameters [ie. the **TLV**, the **PEL**, the **REL**, or the **MAK**], resulting from the <u>combined</u> effects of <u>all</u> the potentially irritating, toxic, or hazardous components in <u>any</u> ambient air system that is being evaluated. To repeat, it is effective for <u>any</u> ambient air system, whether there is only a single volatile or aerosolized component in it, or many such components. In applying this Equation, the individual performing the calculation <u>must</u> know the specific value of the Standard [ie. in the example Formula, the **REL**] that has been established for <u>each</u> of the various components that are contained in the ambient air that is being evaluated. If, for example, the evaluation were to involve an 8-hour **Recommended Exposure Limit – Time Weighted Average** determination for an air mass containing the three different refrigerants: R-12, R-22, & R-112, then the **REL-TWA**s for these three compounds <u>would</u> <u>have</u> <u>to</u> <u>be</u> <u>known</u>; for reference, these three are, respectively: **REL-TWA**$_{R-12}$ = 1,000 ppm, **REL-TWA**$_{R-22}$ = 1,000 ppm, & **REL-TWA**$_{R-112}$ = 500 ppm. The effective Time Weighted Average Concentration values [ie. the **TWA**$_i$'s] in the numerator of Equation **#1B** would be determined by using Equation **#1A**, from Page 3-1.

$$\% REL = 100 \left[\sum_{i=1}^{n} \frac{TWA_i}{REL_i} \right] = 100 \left[\frac{TWA_1}{REL_1} + \frac{TWA_2}{REL_2} + \cdots + \frac{TWA_n}{REL_n} \right]$$

Where: $\% REL$ = the **Effective Percent Exposure Level** from the perspective of the Recommended Exposure Limit Standards that was achieved for the mixture being evaluated, expressed as a percentage;

TWA_i = the **Time Weighted Average Concentration** of the ith component in this mixture, &

REL_i = the listed **Recommended Exposure Limit** (or, in this case, the **REL-TWA**) of the ith component.

Note: If the **% REL** ≤ 100%, then it can be inferred that the **Effective REL** for the mixture, as a whole, <u>has</u> <u>not</u> <u>been</u> exceeded; on the other hand, if this **% REL** > 100%, then the inference is that the **Effective REL** for the mixture <u>has</u> <u>been</u> exceeded.

Equation 1C:

The following Equation, #1C, is used to determine the **Effective Exposure Limit** that exists for <u>any</u> equilibrium vapor phase that is in contact with <u>any</u> well defined liquid mixture containing two or more different volatile components. It can be applied to <u>any</u> of the established parameters [ie. the **TLV**, the **PEL**, the **REL**, or the **MAK**]; the value of the individual Exposure Limit Standard for <u>each</u> of the components in the liquid mixture <u>must be</u> known. If this **Effective Exposure Limit** is determined EITHER by using <u>different</u> Exposure Limit Standards [ie. **TLV-TWAs** coupled with **PEL-TWAs**], OR by using the <u>same</u> Exposure Limit Standard, <u>with</u> <u>different</u> <u>time</u> <u>interval</u> <u>bases</u> [ie. the **REL-STEL**$_{\text{hydrazine}}$ which is evaluated over a 120 minute period, and the **REL-STEL**$_{\text{1,4-dioxane}}$ which is evaluated over a 30 minute period], then the resultant calculated **Effective Exposure Limit** is considered to be <u>neither</u> <u>valid</u> <u>nor</u> <u>accurate</u>. This formula assumes that the overall composition of the vapor phase will be <u>identical</u> to that of the liquid phase, and although this would rarely — if ever — be true, the Equation is considered to be a useful tool for evaluating certain vapor mixtures. This overall Equation was originally proposed for use by the American Conference of Government Industrial Hygienists [the ACGIH]; therefore, it is used most commonly for the determination of this organization's Exposure Limit Standard — namely, the **TLV**.

$$TLV_{\text{effective}} = \left| \frac{1}{\sum\limits_{i=1}^{n} \dfrac{f_i}{TLV_i}} \right| = \left[\frac{1}{\dfrac{f_1}{TLV_1} + \dfrac{f_2}{TLV_2} + \cdots + \dfrac{f_n}{TLV_n}} \right]$$

Where: f_i = the **Weight Fraction** of the ith component in the liquid mixture;

TLV_i = the **Threshold Limit Value** [TLV] – this could be <u>any</u> established Exposure Limit – of the ith component, expressed in **mg/m^3**, <u>NOT</u> **ppm**; &

$TLV_{\text{effective}}$ = the **Effective Threshold Limit Value** — this parameter, too, could be <u>any</u> established Exposure Limit — for the entire mixture of components, also expressed in **mg/m^3**, <u>NOT</u> **ppm**.

Calculations Involving the Conversion of Concentration Units:

Equations 2A & 2B:

The following two Equations, #'s **2A & 2B**, are used to convert sets of **Mass Based Concentration** units to their **Volume Based Concentration** equivalents [e.g. converting from concentrations expressed in **mg/m³** to those expressed in **ppm(vol)**]. The first Equation, **#2A**, would be used to affect this conversion under Normal Temperature and Pressure [NTP] conditions, while the second, **#2B**, would be used under Standard Temperature and Pressure [STP] conditions.

Equation 2A:

$$C_{vol} = \frac{24.45}{MW_i}[C_{mass}] \qquad @ \text{ NTP}$$

Equation 2B:

$$C_{vol} = \frac{22.41}{MW_i}[C_{mass}] \qquad @ \text{ STP}$$

Where: C_{vol} = the **Volume Based Concentration** of the component of interest, namely, the **i**th component, measured in **ppm(vol)**;

C_{mass} = the **Mass Based Concentration** of the same component, namely the **i**th component, in **mg/m³**; &

MW_i = the **Molecular Weight** of this same **i**th component

Equations **2C & 2D**:

The following two Equations, #'s **2C & 2D**, are used to convert sets of **Volume Based Concentration** units to their **Mass Based Concentration** equivalents [e.g. converting from concentrations expressed in **ppm(vol)** to those expressed in **mg/m³**]. The first Equation, **#2C**, would be used to affect this conversion under Normal Temperature and Pressure [**NTP**] conditions, while the second, **#2D**, would be used under Standard Temperature and Pressure [**STP**] conditions.

Equation **2C**:

$$C_{mass} = \frac{MW_i}{24.45}[C_{vol}] \qquad @\ NTP$$

Equation **2D**:

$$C_{mass} = \frac{MW_i}{22.41}[C_{vol}] \qquad @\ STP$$

Where: C_{mass}, C_{vol}, & MW_i are all as defined on Page 3-4

Calculations Involving the TLVs for Free Silica Dust in Air:

<u>Equations 3A, 3B, & 3C</u>:

The following three Equations, #'s **3A**, **3B**, & **3C**, are now used <u>only</u> <u>very</u> <u>infrequently</u>, and <u>only</u> when required to provide a rough approximation of the respective required TLV. These Equations were developed and promulgated by the American Conference of Government Industrial Hygienists [the ACGIH], and are used, therefore, <u>only</u> to determine **Threshold Limit Value Exposure Limits**. In order these three Equations are as follows: **#3A** — For Respirable Quartz Dusts; **#3B** — For Total Dusts; and **#3C** — For Mixtures of the Three Most Common Silica Dusts.

<u>Equation 3A</u>:

$$TLV_{quartz} = \left\lfloor \frac{10\,mg\,/\,m^3}{\%RQ + 2} \right\rfloor$$

Where: TLV_{quartz} = the **8-hour TWA Threshold Limit Value** for respirable quartz dust, measured in **mg/m³**; &

$\%RQ$ = the **Percentage of Respirable Quartz Dusts** in the sample being evaluated, expressed as a percentage.

<u>Equation 3B</u>:

$$TLV_{dust} = \left\lfloor \frac{30\,mg\,/\,m^3}{\%Q + 2} \right\rfloor$$

Where: TLV_{dust} = the **8-hour TWA Threshold Limit Value** for total dusts, measured in **mg/m³**; &

$\%Q$ = the **Percentage of Quartz Dusts** in the sample being evaluated, expressed as a percentage.

Equation **3C**:

$$TLV_{mix} = \left[\frac{10\,mg\,/m^3}{\%Q + [2][\%C] + [2][\%T] + 2} \right]$$

Where: TLV_{mix} = the **8-hour TWA Threshold Limit Value** for the complex mixture of silica dusts, measured in **mg/m³**;

$\%Q$ = the **Percentage of Quartz Dusts** in the sample being evaluated, expressed as a percentage;

$\%C$ = the **Percentage of Cristobalite Dusts** in the sample being evaluated, also expressed as a percentage; &

$\%T$ = the **Percentage of Tridymite Dusts** in the sample being evaluated, this one, also expressed as a percentage.

Laws of Physical Chemistry:

Standard Gas Laws:

The following five Formulae make up the five Standard Gas Laws, which are, in the order in which they will be presented and discussed: **Boyle's Law** (Equation **#4A**); **Charles' Law** (Equation **#4B**); **Gay-Lussac's Law** (Equation **#4C**); the **General Gas Law** (Equation **#4D**); and the **Ideal Gas Law** (Equation **#4E**).

Equation 4A:

The following Equation, **#4A**, is **Boyle's Law,** which is the relationship that de-scribes how the **Pressure** and **Volume** of a gas varies <u>under</u> <u>conditions</u> <u>of</u> <u>constant</u> <u>temperature</u>.

$$P_1 V_1 = P_2 V_2$$

Where:

P_1 = the **Pressure** of a gas @ Time #1, measured in some suitable pressure units;

V_1 = the **Volume** of a gas @ Time #1, measured in some suitable volumetric units;

P_2 = the **Pressure** of the same gas @ Time #2, mea-sured in the <u>same</u> pressure units as P_1 above; &

V_2 = the **Volume** of the same gas @ Time #2, mea-sured in the <u>same</u> volumetric units as V_1 above.

Equation **4B**:

The following Equation, **#4B**, is **Charles' Law**, which is the relationship that describes how the **Volume** and the **Absolute Temperature** of a gas varies <u>under</u> <u>conditions</u> <u>of</u> <u>constant</u> <u>pressure</u>.

$$\frac{V_1}{T_1} = \frac{V_2}{T_2}$$

Where: V_1 & V_2 are each as defined on Page 3-8;

T_1 = the **Absolute** **Temperature** of a gas @ Time #1, measured in either °K or °R; &

T_2 = the **Absolute** **Temperature** of the same gas @ Time #2, measured in the <u>same</u> Absolute Temperature units as T_1 above.

Equation **4C**:

The following Equation, **#4C**, is **Gay-Lussac's Law**, which is the relationship that describes how the **Pressure** and **Temperature** of a gas varies <u>under</u> <u>conditions</u> <u>of</u> <u>constant</u> <u>volume</u>.

$$\frac{P_1}{T_1} = \frac{P_2}{T_2}$$

Where: P_1 & P_2 are each as defined on Page 3-8; &

T_1 & T_2 are each also as defined above on this page.

Equation **4D**:

The following Equation, **#4D**, is the **General Gas Law**, which is the more generalized relationship involving changes in <u>any</u> of the basic measurable characteristics of <u>any</u> gas. This relationship permits the determination of the value of <u>any</u> of the three basic characteristics of the gaseous material being evaluated — namely: its **Pressure**, its **Temperature**, and/or its **Volume** — that might have resulted from changes in one or both of the other two characteristics.

$$\frac{P_1 V_1}{T_1} = \frac{P_2 V_2}{T_2}$$

Where: P_1 & P_2 are each as defined on Page 3-8;

V_1 & V_2 are also each as defined on Page 3-8;

T_1 & T_2 are each also as defined Page 3-9.

Equation **4E**:

The following Equation, **#4E**, is the **Ideal Gas Law**, and is one of the most commonly used Equations of State (it is also frequently called the **Perfect Gas Law**). Like the immediately preceding formula, this one provides the necessary relationship for determining the value of <u>any</u> of the measurable characteristics of a gas — namely, again: its **Pressure**, its **Temperature**, and/or its **Volume**; however, it <u>does</u> <u>not</u> <u>require</u> that one know these characteristics at some alternative state or condition. Unlike Equation **#4D**, from the previous page, it <u>does</u> <u>require</u> that the <u>quantity</u> of the gas involved in the determination be known (ie. the **Number of Moles** involved, or the weight and identity of the gas involved, etc.).

$$PV = nRT$$

Where:

P = the **Pressure** of the gas, in some appropriate units;

V = the **Volume** of the gas, in some appropriate units;

n = the **Number of Moles** of a gas

$= \dfrac{\text{weight of the gas in grams}}{\text{molecular weight of the gas}}$;

R = the **Universal Gas Constant**, in units appropriate to those used for the pressure, volume, and temperature:

$= 0.0821 \dfrac{[\text{liter}][\text{atmosphere}]}{[°K][\text{gram mole}]}$;

$= 62.36 \dfrac{[\text{liter}][\text{mm Hg}]}{[°K][\text{gram mole}]}$;

$= 0.00359 \dfrac{[\text{cubic feet}][\text{millibar}]}{[°R][\text{pound mole}]}$; &

T = the **Absolute Temperature** of the gas, measured in some appropriate Absolute Temperature units.

The Mole

Equation 5A:

The following Equation, #5A, involves the Mass Based relationship that defines the **Mole**. Specifically, a **Mole** (which is equivalent to a **Gram Mole** or a **Gram Molecular Weight**) of any compound or chemical is that quantity, which when weighed will have a weight, IN GRAMS, that is <u>numerically</u> <u>equivalent</u> to that compound or chemical's Molecular Weight, expressed in Atomic Mass Units [amu's]. Analogously, the **Pound Mole** of any compound or chemical is that quantity, which when weighed will have a weight, IN POUNDS, that is <u>numerically</u> <u>equivalent</u> to that compound or chemical's Molecular Weight, expressed in Atomic Mass Units [amu's]. The **Pound Mole** is much less commonly used than is the **Gram Mole** — thus, as implied above, when a reference identifies only "a Mole", it can be reasonably assumed that it is the **Gram Mole** that is involved. A **Pound Mole** will have a mass that is 453.59 times greater than that of a **Gram Mole**.

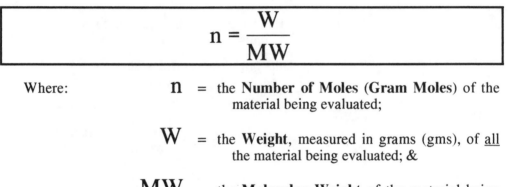

$$n = \frac{W}{MW}$$

Where:
n = the **Number of Moles (Gram Moles)** of the material being evaluated;

W = the **Weight**, measured in grams (gms), of <u>all</u> the material being evaluated; &

MW = the **Molecular Weight** of the material being evaluated, expressed in Atomic Mass Units (amu's).

Equation **5B**:

The following Equation, **#5B**, involves the Numerical Count Based relationship that also defines the **Mole**. A **Mole** (= a **Gram Mole** = a **Gram Molecular Weight**) of <u>any</u> compound or chemical will <u>always</u> <u>contain</u> **Avogadro's Number**, usually identified as, N_0, which equals — numerically — 6.022 x 10^{23} molecules.

$$n = \frac{Q}{N_A} = \frac{Q}{6.022 \times 10^{23}}$$

Where: n is as defined on Page 3-12;

             Q = the actual count or **Number of Molecules** of the material being evaluated; &

             N_A = **Avogadro's Number** = 6.022 x 10^{23}.

<u>Equations **5C**, & **5D**</u>:

The following Equations, #'s **5C** & **5D**, involve the third of the three relationships, these based on Volumetric considerations, that define the **Mole**. Again, a **Mole** (= a **Gram Mole** = a **Gram Molecular Weight**) of <u>any</u> compound or chemical <u>that exists as a gas</u> under the conditions of Pressure and Temperature under which the evaluation is being made will have a very specific and predictable **Volume**. Under **STP** Conditions, the **Molar Volume]$_{STP}$** = 22.414 liters; under NTP Conditions, the **Molar Volume]$_{NTP}$** = 24.465 liters.

Equation **5C**:

$$n = \frac{V}{V_{molar-STP}} = \frac{V}{22.414} \quad @ \ \textbf{STP}$$

Equation **5D**:

$$n = \frac{V}{V_{molar-NTP}} = \frac{V}{24.465} \quad @NTP$$

Where: n is as defined on Page 3-12;

V = the **Volume** of the gaseous material being evaluated, measured in liters;

$V_{molar-STP}$ = the **Molar Volume]$_{STP}$** at Standard Temperature and Pressure Conditions, expressed in liters and having a value of **22.414 liters**; &

$V_{molar-NTP}$ = the **Molar Volume]$_{NTP}$** at Normal Temperature and Pressure Conditions, expressed in liters and having a value of **24.465 liters**.

The Gas Density Law:

<u>Equation 6A</u>:

The following Equation, **#6A**, provides a vehicle for relating the **Density** of <u>any</u> gas to its **Pressure** and **Temperature** characteristics.

$$\frac{\rho_1 T_1}{P_1} = \frac{\rho_2 T_2}{P_2}$$

Where: P_1 & P_2 are each as defined on Page 3-8;

T_1 & T_2 are each also as defined on Page 3-8.

ρ_1 = the **Density** of the gas @ Time #1, measured most commonly, in units of grams/cubic centimeter (gms/cm^3); &

ρ_2 = the **Density** of the gas @ Time #2, in the same units as ρ_1.

Dalton's Law of Partial Pressures:

Equation 7A & 7B:

The following two Equations, #'s **7A** & **7B**, are expressions of **Dalton's Law of Partial Pressures**; in general, this Law simply states that the **Total Pressure** exerted by the mixture of different gases in any gas volume will be, and is, equal to the Sum of the **Partial Pressures** of each individual component in the overall gas mixture. **Dalton's Law** also states that each individual component in any gas mixture exerts its own individual **Partial Pressure**, in the exact ratio as its **Mole Fraction** (or volume based **Concentration**) in that mixture.

Equation 7A:

$$P_{total} = \sum_{i=1}^{n} P_i = P_1 + P_2 + P_3 + \cdots + P_n$$

Where: P_{total} = the **Total Pressure** of a gas mixture containing "**n**" different identifiable gaseous components, measured usually in millimeters of Mercury (mm Hg); &

P_i = the **Partial Pressure** of the **i**th member of the "**n**" total components of the entire gas mixture, measured in the same units as P_{total} .

The second portion of the overall statement of **Dalton's Law** involves the previously mentioned relationship of the **Partial Pressure** of any component to its **Mole Fraction** in the entire mixture. This relationship also provides a way to determine the volume based **Concentration** of any individual gaseous component in a mixture of several other gases OR in air.

Equation **7B**:

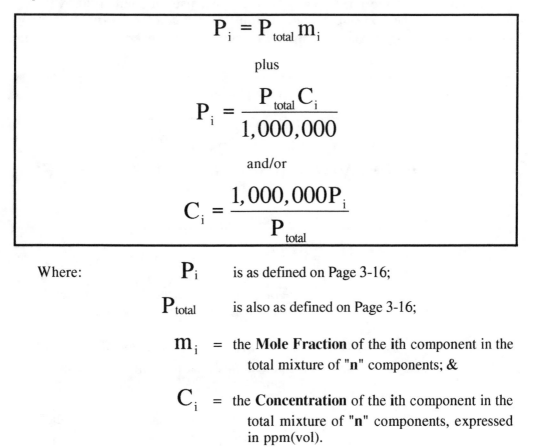

$$P_i = P_{total}\, m_i$$

plus

$$P_i = \frac{P_{total}\, C_i}{1,000,000}$$

and/or

$$C_i = \frac{1,000,000 P_i}{P_{total}}$$

Where: P_i is as defined on Page 3-16;

P_{total} is also as defined on Page 3-16;

m_i =   the **Mole Fraction** of the ith component in the total mixture of "**n**" components; &

C_i =   the **Concentration** of the ith component in the total mixture of "**n**" components, expressed in ppm(vol).

Raoult's Law

Equation 7C:

The following Equation, **#7C**, is known as **Raoult's Law**. This Law involves the relationship between the **Partial Vapor Pressure** of each of the components in a Solution containing two or more volatile components (ie. the Partial Pressure of each component in the mixed Vapor Phase that exists in equilibrium with the Solution), AND the **Solution Mole Fraction** of each component (not the Vapor Phase Mole Fraction!!!) and the **Vapor Pressure** it would exert if it was alone in the pure liquid state.

$$PVP_i = m_i VP_i \quad \text{for the ith component in the solution}$$

$$\text{and}$$

$$TVP_{solution} = \sum_{i=1}^{n} m_i VP_i = m_1 VP_1 + m_2 VP_2 + \cdots + m_n VP_n$$

for the solution as a whole

Where: PVP_i = the **Partial Vapor Pressure** exerted by the ith component in a solution of "**n**" different components, measured most frequently in millimeters of mercury (mm Hg);

m_i = the **Solution Mole Fraction** of the ith component in the solution of "**n**" different components;

VP_i = the **Vapor Pressure** of the "pure liquid" ith component of the solution, always listed as a function of its temperature, almost always measured in millimeters of mercury (mm Hg); &

$TVP_{solution}$ = the **Total Vapor Pressure** for the solution as a whole, also measured in millimeters of mercury (mm Hg).

Settling Velocity of Various Types of Particles Suspended in the Air

<u>Equation 8A</u>:

This Equation, **#8A**, is the most commonly used approximation to **Stoke's Law**. It relates the **Terminal**, or **Settling Velocity**, <u>in air</u> of any suspended particle (ie. dusts, fumes, mists, etc.) to that particle's **Effective Diameter** and **Density**. In <u>every</u> <u>case</u>, the **Density** of the settling particle <u>must</u> <u>be</u> <u>greater</u> than the Density of the air – which is 0.00129 grams/cm^3 at **STP**, and/or 0.00118 grams/cm^3 at **NTP** – in order for the particle to settle at all. Particles less dense than this are <u>ex-</u><u>tremely</u> <u>rare</u>, and will <u>never</u> settle in air. At the other extreme, particles with an **Effective Diameter** <u>greater</u> than 50 - 60 microns will simply "fall" to earth, rather than "settle," thus this approximation to **Stoke's Law** (Equation **#8A**) should <u>never</u> be used in any case where the **Effective Diameter** of the particle being considered is at or above this size range.

$$V_s = 0.003 \left[\frac{\rho_{particle}}{\rho_{air-STP}} \right] d_{particle}^2 = 2.335 \left[\rho_{particle} \, d_{particle}^2 \right] @ \text{ STP}$$

$$V_s = 0.003 \left[\frac{\rho_{particle}}{\rho_{air-NTP}} \right] d_{particle}^2 = 2.548 \left[\rho_{particle} \, d_{particle}^2 \right] @ \text{ NTP}$$

Where:

V_s = the **Settling Velocity** in air of the particle being considered, measured in cm/sec;

$\rho_{particle}$ = the **Density** of the particle being considered, measured in gms/cm^3;

$\rho_{air-STP}$ = the **Density** of air under **STP** conditions, identified as 0.00129 gms/cm^3;

$\rho_{air-NTP}$ = the **Density** of air under **NTP** conditions, identified as 0.00118 gms/cm^3; &

d = the **Effective Diameter** of the particle, measured in microns, (μ).

Ionizing and Non-Ionizing Radiation:

Frequency (Wavenumber) and Wavelength Relationships:

Equation 9A:

The following Equation, #9A, is the very basic relationship, the application of which will permit the determination of the **Wavelength (Wavenumber)** that corresponds to a known **Frequency** OR the **Frequency** that corresponds to a known **Wavelength (Wavenumber)**. This relationship applies to all areas of the Electromagnetic Spectrum — ranging from the very long **Wavelength** [low **Frequency**] Non-Ionizing Radiation sections of this Spectrum, to the very short **Wavelength** [high **Frequency**] Ionizing Radiation sections of this same Spectrum.

$$C = f\lambda \quad \text{or} \quad \lambda = \frac{C}{f} \quad \text{or} \quad f = \frac{C}{\lambda}$$

Where:

C = the **Speed of Light**, in this case, usually measured in cm/sec, and having a value of 3×10^{10} cm/sec;

f = the **Frequency** of any photon of Electromagnetic Radiation, measured in cycles/sec (or Hertz = Hz); &

λ = the **Wavelength** of any photon of Electromagnetic Radiation, measured in centimeters/wave, or occasionally microns/wave (microns/wave = μ/wave).

Equation **9B**:

The following Equation, **#9B**, simply identifies the relationship between the **Wavelength** of any photon in the Electromagnetic Spectrum, and its **Wavenumber**.

$$\lambda = \frac{1}{\nu} \quad \text{or} \quad \nu = \frac{1}{\lambda}$$

Where:

λ is as defined on Page 3-20; &

ν = the **Wavenumber** of the photon, measured in waves/unit length (usually, waves/cm or just cm^{-1}).

Radioactive Decay:

Equation **10A**:

The following Equation, **#10A**, identifies the **Quantity** of <u>any</u> radioactive element that would present in a sample of this material (ie. the **Quantity** of that element that has <u>not</u> <u>yet</u> undergone radioactive disintegration), at any incremental **Time Period** after an initial **Quantity** has been determined and identified. With radioactive decay, the Number of Disintegrations per Unit Time will be proportional to BOTH the **Radioactive Decay Constant** for that nuclide, AND the actual **Numeric Count** of Nuclei of that element (ie. the **Quantity**) that remain undisintegrated in the sample.

$$N_t = N_0 e^{-kt}$$

Where:

N_t = the **Quantity** of any radioactive element present at any time, "**t**", usually measured <u>either</u> in mass units (mg or μg), <u>OR</u> as a specific **Numeric Count** of the undisintegrated Nuclei remaining in the sample;

N_0 = the **Initial Quantity** of that same radioactive element that was present at the time **t = t₀** (0 seconds, 0 minutes, 0 hours, 0 days, or <u>whatever</u> unit of time is appropriate to the units in which the **Radioactive Decay Constant** has been expressed) — this is the "Starting" or **Initial Quantity** of this element, always expressed in the same units as N_t above;

k = the **Radioactive Decay Constant**, measured in reciprocal units of time (ie. /seconds, /minutes, /hours, /days, or /years, etc.); &

t = the **Time Interval** that has passed since the **Initial Quantity** of material was determined, expressed in an appropriate unit of time; care must be exercised to make certain that the units of "k" and "t" are <u>consistent</u> with each other.

Equation **10B**:

The following Equation, **#10B**, relates the frequently used and commonly listed **Half Life** of a radioactive element to its less common **Radioactive Decay Constant**.

$$T_{1/2} = \frac{0.693}{k} \quad \text{and/or} \quad k = \frac{0.693}{T_{1/2}}$$

Where: $T_{1/2}$ = the **Half Life** of the radioactive element under consideration; this parameter is most commonly expressed in those units of time that are the reciprocal of the units that are used for the **Radioactive Decay Constant**;

k is as defined on Page 3-22.

Equation **10C**:

The following Equation, **#10C**, provides a relationship with which the number of **Radioactive Decays per Unit Time** or **Disintegrations per Unit Time** can be determined, if <u>either</u> the **Radioactive Decay Constant** <u>or</u> the **Half Life** of the element under consideration is known.

$$A_t = k N_0 e^{-kt} = \frac{0.693}{T_{1/2}} N_0 e^{-(0.693)t/T_{1/2}}$$

Where:

A_t = the **Number of Disintegrations per Unit Time**, that will be occurring at any time, **t**. The units for this parameter will be the same as those of the **Half Life** (or the reciprocal of the units of the **Radioactive Decay Constant**);

k is as defined on Page 3-22;

N_0 is also as defined on Page 3-22;

$T_{1/2}$ is as defined on Page 3-23;

t finally, is also as defined on Page 3-22.

Dose And/or Exposure Calculations:

<u>Equation 11A</u>:

The following Equation, #11A, is applicable <u>only</u> to **Dose Exposure Rates** caused by high energy X-Rays and/or γ-Rays (as well as — hypothetically, but certainly not practically — <u>any</u> <u>other</u> Electromagnetic Radiation of still shorter wavelengths, such as Cosmic Rays). Determinations of these **Dose Exposure Rates** are, to all extents and purposes, limited to medical applications. In order to be able to determine these **Dose Exposure Rates**, one must know some <u>very</u> <u>specific</u> and <u>unique</u> source based radiological data (ie. the **Radiation Constant** of the source), as well as the **Radiation Source Activity** and the **Distance** from the source to the point at which **Dose Exposure Rate** measurements are to be made.

$$E = \frac{\Gamma A}{d^2}$$

Where:

E = the **Dose Exposure Rate** that has resulted from an individual's exposure to some specific X- or γ-Ray radiation source, for which the specific **Radiation Constant**, Γ, is known;

Γ = the **Radiation Constant** for the X- or γ-Ray radioactive nuclide being considered, expressed in units of Rad·cm^2/mCi·hr;

A = the **Radiation Source Activity**, measured usually in millicuries (mCi's); &

d = the **Distance**, measured in centimeters, of the "Target" from the radiation source.

Equation 11B:

This Equation, **#11B**, provides for the conversion of the **Physical Radiation Doses**, expressed in either **Rads** or **Grays**, to a more useful form (useful, from the perspective of the overall impact of the specific Type of **Radiation Dose** on the individual who has been exposed). This alternative and more useful form of **Radiation Dose** is expressed either in **Rems** or in **Sieverts**, both of which measure the "Relative Hazard" caused by the energy transfer that results from an individual's exposure to various different Types of Radiation. To make these determinations, a "**Quality Factor**" is used to adjust the **Rads** or **Grays** Dose into its **Rems** or **Sieverts** equivalent. This **Quality Factor [QF]** is a simple multiplier that adjusts for the effective *Linear Energy Transfer* (*LET*) that is produced on a target by each Type of Radiation. The higher the *LET*, the greater will be the damage that can be caused by the Type of Radiation being considered; thus, this alternative **Dose Factor** measures the overall biological effect, or impact, of an otherwise "simply" measured **Radiation Dose**.

$$D_{Rem} = D_{Rad}[QF] \quad \& \quad D_{Sieverts} = D_{Grays}[QF]$$

Where: D_{Rem} or $D_{Sieverts}$ = the adjusted **Radiation Dose** in the more useful "effect related" form, measured in either Rems or *Sieverts [SI Units]*;

D_{Rad} or D_{Grays} = the **Physical Radiation Dose** — & independent of the type of radiation, measured in either Rads or *Grays [SI Units]*; &

QF = the **Quality Factor**, which is a properly dimensioned factor — EITHER Rems/Rad OR Sieverts/Gray, as applicable — that is, itself, a function of the Type of Radiation being considered [see the following Tabulation].

Tabulation of Quality Factors [QFs] by Radiation Type

Types of Radiation	Quality Factors — QFs
X-Rays <u>or</u> γ-Rays	1.0
β-Rays [positrons <u>or</u> electrons]	1.0
Thermal Neutrons	5.0
Slow Neutrons	4.0 - 22.0
Fast Neutrons	3.0 - 5.0
Heavy, Charged Particles [Alphas, etc.]	20.0

Calculations Involving the Reduction of Radiation Intensity Levels:

<u>Equation 12A</u>:

This Equation, #12A, identifies the effect that Shielding Materials have in reducing the Intensity Levels of ionizing radiation. **Radiation Emission Rates** can be reduced by interposing Shielding Materials between the radiation source and the receptor, <u>or</u> by increasing the source-to-receptor distance <u>OR</u> by using both approaches simultaneously. The approach represented by this formula deals <u>solely</u> with the use of Shielding Materials, and involves the **Half-Value Layer (HVL)** concept. A **Half Value Layer** represents the **Thickness** of <u>any</u> Shielding Material that would reduce, by one half, the Intensity Level of incident γ-radiation.

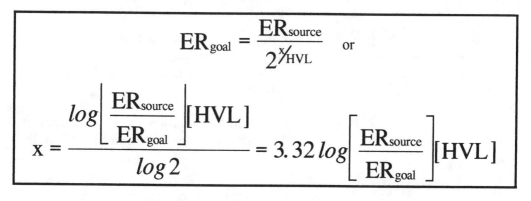

$$ER_{goal} = \frac{ER_{source}}{2^{x/HVL}} \quad or$$

$$x = \frac{log\left[\dfrac{ER_{source}}{ER_{goal}}\right][HVL]}{log\,2} = 3.32\,log\left[\dfrac{ER_{source}}{ER_{goal}}\right][HVL]$$

Where: ER_{goal} = the target **Radiation Emission Rate**, measured in units of radiation dose per unit time, (Rads/hour);

ER_{source} = the observed **Radiation Emission Rate** to be reduced by interposing Shielding Materials; measured in the same units as ER_{goal};

X = the **Thickness** of Shielding Material required to reduce the measured **Radiation Emission Rate** to the level desired, usually measured in inches or centimeters; &

HVL = the **Half Value Thickness** of the Shielding Material being evaluated (ie. the **Thickness** of this material that will halve the Intensity Level of incident γ-radiation), measured in the same units as "X", above.

<u>Equation **12B**</u>:

The following Equation, **#12B**, is the relationship that describes the effect of increasing the **Distance** between a <u>point source</u> of γ-radiation and a receptor, as an alternative method for decreasing the incident **Radiation Intensity** on the receptor, as measured by the effective **Radiation Emission Rate**. The relationship involved is basically geometric, and is frequently described as **The Inverse Square Law**.

$$ER_A \, S_A^2 = ER_B \, S_B^2 \quad \text{or} \quad \frac{ER_A}{ER_B} = \frac{S_B^2}{S_A^2}$$

Where:

ER_A = the **Radiation Emission Rate**, or **Radiation Intensity**, measured at distance, "**A**", in units of radiation dose per unit time, ie. Sieverts/hour;

ER_B = the **Radiation Emission Rate**, or **Radiation Intensity**, measured at distance, "**B**", in the same units as, ER_A, above;

S_A = the "**A**" **Distance**, or the distance between the Source and the "**A**" Receptor, measured in some appropriate unit of length; &

S_B = the "**B**" **Distance**, or the distance between the Source and the "**B**" Receptor, measured in the same length units as, S_A, above.

Optical Density Calculations:

<u>Equation 13A</u>:

The following Equation, #13A, describes the relationship between the absorption of monochromatic visible light [laser light], and the length of the path this beam of light follows through some absorbing medium. This formula relies on the fact that <u>each</u> incremental thickness of this absorbing medium will absorb the same equal fraction of the incident radiation as will each other incremental thickness of this same medium. The ratio of the **Incident Beam Intensity** to the **Transmitted Beam Intensity** is used to provide the Factor that is necessary to determine the **Optical Density** of the medium. This relationship is routinely used to determine the <u>intensity</u> <u>diminishing</u> capabilities, or the **Optical Density**, of the protective goggles that must be worn by individuals who must operate equipment that makes use of high intensity lasers as an integral operating part.

$$OD = log \left[\frac{I_{incident}}{I_{transmitted}} \right]$$

Where:

OD = the measured **Optical Density** of the material being evaluated, this parameter is dimensionless;

$I_{incident}$ = the **Incident Laser Beam Intensity**, measured in units of power/unit area (W/cm^2); &

$I_{transmitted}$ = the **Transmitted Laser Beam Intensity**, measured in the same units as $I_{incident}$.

Microwave Calculations Involving the *Near* and the *Far Fields*

Equation 14A:

The following Equation, **#14A**, provides the necessary relationship for determining the **Distance to the *Far Field*** for any radiating circular microwave antenna. The *Far Field* is that region that is sufficiently distant from the radiating source antenna, that one may consider that there is no longer any interaction between the source of the microwave radiation, and the electromagnetic waves being produced by this source. The *Near Field* is every portion of the radiation field that is not included in the *Far Field*., ie. it is that area that is closer to the source antenna than is the *Far Field*.

$$r_{FF} = \frac{A}{2\lambda} = \frac{\pi D^2}{8\lambda}$$

Where:

r_{FF} = the **Distance to the *Far Field*** from the microwave radiating antenna (all distances equal to or greater than r_{FF} are considered to be in the *Far Field*; all distances less than this value, in the *Near Field*), these distances are usually measured in centimeters;

A = the **Area** of the radiating circular antenna, measured in square centimeters (cm^2)

$$= \frac{\pi D^2}{4};$$

D = the circular microwave antenna **Diameter**, measured in centimeters (cm); &

λ = the **Wavelength** of microwave energy being radiated by the antenna, measured in centimeters (cm).

<u>Equation **14B**</u>:

The following Equation, **#14B**, provides the relationship for determining the **Microwave Power Density** levels in the *Near Field* that are produced by a microwave antenna, radiating at a known **Average Power Output**.

$$W_{NF} = \frac{4P}{A} = \frac{16P}{\pi D^2}$$

Where:

W_{NF} = the *Near Field* **Microwave Power Density**, measured in milliwatts/cm^2 (mw/cm^2);

P = the **Average Power Output** of the microwave radiating antenna, measured in milliwatts (mw);

A is as defined on Page 3-30; &

D is also as defined on Page 3-30.

Equations **14C, & 14D**:

The following two Equations, #'s **11C & 11D**, provide the basic approximate relationships that are used for calculating either microwave **Power Density Levels** in the *Far Field* (#14C); OR alternatively, for determining the actual *Far Field* **Distance** from a radiating microwave antenna at which one would expect to find some specific **Power Density Level** (#14D). Unlike the previous Equation (ie. #14B), these two formulae are empirical; however, they may both be regarded as sources of reasonably accurate values for the **Power Density Levels** at points in the *Far Field* (#14C), or for various *Far Field* **Distances** (#14D).

Equation **14C**:

$$W_{FF} = \frac{AP}{\lambda^2 r^2} = \frac{\pi D^2 P}{4 \lambda^2 r^2}$$

Equation **14D**:

$$r = \frac{1}{\lambda} \sqrt{\frac{AP}{W_{FF}}} = \frac{D}{2\lambda} \sqrt{\frac{\pi P}{W_{FF}}}$$

Where: W_{FF} = the microwave **Power Density Level** at a point in the "Far Field" that is "r" centimeters distant from the circular antenna, measured in milliwatts/cm^2 (mw/cm^2);

r = the *Far Field* **Distance** (from the point where the **Power Density Level** is being evaluated) to the radiating circular antenna, measured in centimeters (cm);

D is as defined on Page 3-30;

A is also as defined on Page 3-30;

λ is also as defined on Page 3-30; &

P is as defined on Page 3-31.

Heat and Cold Stress Related:

Heat & Cold Stress (Indoor/Outdoor), with & without Solar Load:

Equation 15A:

The following Equation, #15A, is the relationship that provides the **Wet Bulb Globe Temperature Index** [**WBGT**] that is applicable only to situations for which there is no solar load (ie. no direct solar input to the condition or circumstance of the area or space being evaluated). Obviously, most indoor situations fulfill this requirement; in addition, any outdoor circumstance wherein the sun has been shaded — and in which, it makes no radiant contribution to the thermal or temperature environment — also fulfills this condition. This category of the **Wet Bulb Globe Temperature Index** is usually identified with an "$_{Inside}$" subscript.

$$ WBGT_{Inside} = 0.7[NWB] + 0.3[GT] $$

Where: $WBGT_{Inside}$ = the **Wet Bulb Globe Temperature Index**, applicable to any situation in which there is no Solar Load, usually measured in °C;

NWB = the **Natural Wet-Bulb Temperature**, usually also measured in °C; however, it should be noted that any temperature scale may be used for these parameters, so long as the units of every temperature parameter in the formula is consistent with the units of every other temperature parameter; &

GT = the **Globe Temperature**, also in units consistent with every other parameter in this formula.

Equation **15B**:

The following Equation, **#15B**, is the alternative relationship that provides the **Wet Bulb Globe Temperature Index [WBGT]** that is applicable to situations for which there <u>is</u> a measurable solar load. Outdoor conditions usually require this approach; and, correspondingly, this category of the **Wet Bulb Globe Temperature Index** is usually identified with an "Outside" subscript.

$$WBGT_{Outside} = 0.7[NWB] + 0.2[GT] + 0.1[DB]$$

Where: $WBGT_{Outside}$ is precisely as defined, under the designation of $WBGT_{Inside}$, on Page 3-33;

NWB is also as defined on Page 3-33;

GT is also as defined on Page 3-33; &

DB = the **Dry-Bulb Temperature**, usually measured in °C; however, it should be noted for this formula, too, that <u>any</u> temperature scale may be used for these parameters, so long as the units of <u>every</u> temperature parameter in the formula is consistent with the units of <u>every other</u> temperature parameter.

Calculations involving WBGT Time Weighted Averages

Equation 16A:

The following Equation, **#16A**, provides the relationship necessary for the determination of the **Effective Time Weighted Average WBGT Index** that represents an average exposure over time, at various different **WBGT Indices**. This formula is directly analogous to every other formula that is used to determine a Time Weighted Average.

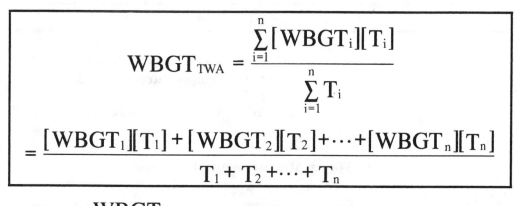

$$WBGT_{TWA} = \frac{\sum\limits_{i=1}^{n}[WBGT_i][T_i]}{\sum\limits_{i=1}^{n}T_i}$$

$$= \frac{[WBGT_1][T_1]+[WBGT_2][T_2]+\cdots+[WBGT_n][T_n]}{T_1+T_2+\cdots+T_n}$$

Where: $WBGT_{TWA}$ = the **Effective Time Weighted Average WBGT Index** that corresponds to a varied thermal exposure over time, usually measured in °C;

$WBGT_i$ = the ith **Wet Bulb Globe Temperature Index** (either "Indoor" or "Outdoor", but NOT mixed "Indoor" and "Outdoor") that was measured over the Time Interval, T_i, usually measured in °C; &

T_i = the ith **Time Interval**, usually measured in hours; however, it can be measured in any useful and consistent units.

Sound and Noise Related:

Approximate Velocity of Sound in Air:

Equation 17A:

This Equation, #17A, is empirical and can be used to calculate accurately, the **Velocity of Sound in Air** — as a function of the **Ambient Air Temperature**.

$$V = 49\sqrt{t + 459}$$

Where:

V = the **Velocity of Sound in Air**, at an **Ambient Air Temperature** of $t\,°F$; &

t = the **Ambient Air Temperature**, measured in °F.

Basic Sound Measurement Definitions:

Equation 18A:

The following Equation, #18A, constitutes the **Definition** of a **Sound Pressure Level**, relating the measured **Analog Sound Pressure Level** to a "Base Reference Analog Sound Pressure Level". Because of the extremely wide variability in these analog values, this parameter is measured in decibels, a logarithmic function better suited to values varying over several orders of magnitude.

$$L_P = 20\,log\left[\frac{P}{P_0}\right] = 20\,log\left[\frac{P}{2 \times 10^{-5}}\right] = 20\,log\,P + 93.98$$

Where:

L_P = the **Sound Pressure Level**, measured in decibels (dB);

P = the measured **Analog Sound Pressure Level**, in units of newtons/square meter (nt/m^2); &

P_0 = the "**Base Reference Analog Sound Pressure Level**", which has been set to equal $2 \times 10^{-5}\,nt/m^2$.

Equation **18B**:

This Equation, **#18B**, constitutes the **Definition** of a **Sound Intensity Level**, relating the measured **Analog Sound Intensity Level** to a "Base Reference Analog Sound Intensity Level". Like the preceding relationship, this one too provides **Sound Intensity Levels** in decibels, and for the same reason.

$$L_I = 10\,log\left[\frac{I}{I_0}\right] = 10\,log\left[\frac{I}{10^{-12}}\right] = 10\,log\,I + 120$$

Where: L_I = the **Sound Intensity Level**, in decibels (dB);

I = the measured **Analog Sound Intensity Level**, in watts/square meter (wts/m^2); &

I_0 = the "**Base Reference Analog Sound Intensity Level**", which has been set to equal 10^{-12} wts/m^2.

Equation **18C**:

The following Equation, **#18C**, constitutes the Definition of a **Sound Power Level**, relating the measured **Analog Sound Power Level** to a "Base Reference Analog Sound Power Level". Like the two preceding relationships, this one, too, provides **Sound Power Levels** in decibels, and for the same reason.

$$L_P = 10\,log\left[\frac{P}{P_0}\right] = 10\,log\left[\frac{P}{10^{-12}}\right] = 10\,log\,P + 120$$

Where: L_P = the **Sound Power Level**, in decibels (dB);

P = the measured **Analog Sound Power Level**, in units of watts (wts); &

P_0 = the "**Base Reference Analog Sound Power Level**", which has been set to equal 10^{-12} wts.

Sound Pressure Levels of Noise Sources in a Free Field:

Equation 19A:

The following Equation, #19A, identifies and relates the specific factors that <u>must</u> be accounted for when one determines the **Effective Sound Pressure Level** of any noise source in a "*Free Field*". For reference, a "*Free Field*" is <u>any</u> region (within which the noise source is located) that can be characterized as being free or void of any and all objects other than the noise source itself. Such a region permits the <u>unhindered</u> <u>propagation</u> of sound from the source in <u>ALL</u> directions. Because noise sources in the real world are seldom, if ever, in a true "*Free Field*", this Equation has, as its final additive factor, a logarithmic term that effectively adjusts the resultant **Sound Pressure Level** for any "*Non-Free Field*" asymmetry factor that has effectively modified or mitigated the "*Free Field*ness" of the region where the noise source is located. As an example, a bell mounted on a wall would not be able to radiate sound in the direction of the wall; rather, it would effectively radiate sound <u>only</u> into a single spatial hemisphere. The factor that is used to achieve this result modification is called the **Directionality Factor**, and is defined below.

$$L_{P-Effective} = L_{P-Source} - 20\,log\,r - 0.5 + 10\,log\,Q$$

Where: $L_{P-Effective}$ = the **Effective Sound Pressure Level**, evaluated at a point that is "r" feet distant from the noise source itself, in decibels (dBs);

$L_{P-Source}$ = the source **Sound Pressure Level**, in decibels (dBs);

r = the **Distance** from the point where the **Effective Sound Pressure Level** is measured, to the noise source, in feet (ft); &

Q = the **Directionality Factor**, a dimensionless parameter — as defined and valued below:

Q = **1** for "spherical omnidirectional" radiating sources

Q = **2** for "single hemisphere" radiating sources

Q = **4** for "single quadrant" radiating sources

Q = **8** for "single octant" radiating sources

Addition of Sound Pressure Levels from Several Independent Sources

Equation 20A:

The following Equation, #20A, is one of the most useful and frequently employed relationships in all of acoustical engineering. It provides the basic methodology for determining the cumulative effect of several noise sources (each, producing noise at an identifiable **Sound Pressure Level**), on an observer. This relationship enables one to determine an **Effective Sound Pressure Level** for that observer. For this determination, we assume the perspective of an observer whose relative location — among two or more noise sources — causes him or her to experience an overall noise exposure that is: (1) obviously greater than would have been the situation involving only a single noise source, but (2) certainly not determinable by simply adding up the measured Sound Pressure Levels of the several noise sources. As a simple and demonstrative example, consider that a Drag Race Starter standing halfway between two dragsters that are each producing sound at 130 dB, does not experience sound at 260 dB.

$$L_{Total} = 10\,log\left[\sum_{i=1}^{n} 10^{L_i/10}\right] \quad or$$

$$L_{Total} = 10\,log\left[10^{L_1/10} + 10^{L_2/10} + \cdots + 10^{L_n/10}\right]$$

Where:

L_{Total} = the total **Effective Sound Pressure Level** resulting from the "**n**" different noise sources, this parameter measured in decibels (dBs); &

L_i = the **Sound Pressure Level** of the **i**th of "**n**" different noise sources, measured in decibels (dBs).

Calculations Involving Sound Pressure Level "Doses"

<u>Equation 21A</u>:

The following Equation, **#21A**, provides the **Maximum Time Period** during which any worker may be exposed to some specifically quantified and/or **Equivalent Sound Pressure Level** from any (or many) noise source(s).

$$T_{max} = \frac{8}{2^{[L-90]/5}}$$

Where:

T_{max} = the **Maximum Time Period** — at any **Equivalent Sound Pressure Level**, L — to which a worker may be exposed during a normal 8-hour workday, measured in some convenient time unit, usually hours; &

L = the **Sound Pressure Level** (or **Equivalent Sound Pressure Level**) being evaluated for this situation, measured in decibels (dBs).

Equation **21B**:

The following Equation, **#21B**, involves the formula that provides the **Effective Daily Dose** that an individual would have experienced as a result of his or her having been exposed to several different **Sound Pressure Levels**, each of these different **SPLs**, for some specific **Time Period** or **Time Interval**.

$$D = \sum_{i=1}^{n} \frac{C_i}{T_{max_i}} = \frac{C_1}{T_{max_1}} + \frac{C_2}{T_{max_2}} + \cdots + \frac{C_n}{T_{max_n}}$$

Where:

D = the **Effective Daily Dose** (of noise), that an individual who has been exposed to the sounds at "**n**" different **Sound Pressure Levels**, each identified exposure lasting for a specific **Time Period** or **Time Interval**, C_i; this parameter, D, is a dimensionless decimal number;

C_i = the overall ith **Time Interval** or **Time Period** during which the individual being considered was exposed to the ith **Sound Pressure Level**; these **Time Intervals** will always have to be measured in some consistent unit of time, usually in hours; &

T_{max_i} = the **Maximum Time Period** that is permitted for the ith specific **Sound Pressure Level** to which an individual could be exposed; as defined by Equation **21A**, on Page 3-40.

Note: For any value of the **Effective Daily Dose**, $D \leq 1.00$, the individual who has experienced this dose level will have accumulated neither an excessive nor a harmful amount of noise. On the other hand, if $D > 1.00$, then the exposure would be classified as potentially harmful.

Equation **21C**:

The following Equation, **#21C**, provides the relationship for determining the **Equivalent Sound Pressure Level** that corresponds to any identified **Daily Dose**.

$$L_{equivalent} = 90 + 16.61 \log D$$

Where: $L_{equivalent}$ = the **Equivalent Sound Pressure Level** that corresponds to any **Daily Dose**, measured in decibels (dBs); &

D = the **Daily Dose**, as defined on the previous page, by Equation **#21B**, namely, on Page 3-41.

Definitions of the A-Scale and of Various Types of Octave Bands

Definition **22A**:

The following Tabulation, as the defining descriptor of the **A-Weighted Sound Scale** (Definition **#22A**) gives both the Deductions and the Increments that must be applied to each of the **Standard Unitary Octave Bands**, in order to make the measured **SPL** measurement correspond to the approximate acoustic response of the human ear. Each of these Deductions and/or Increments would be applied to the true **Equivalent Sound Pressure Level**, that was measured for the specific full **Unitary Octave Band** being considered.

Octave Band, in Hertz	Deduction, in dB	Increment, in dB
31 Hz	39 dB	
63 Hz	26 dB	
125 Hz	16 dB	
250 Hz	9 dB	
500 Hz	3 dB	
1,000 Hz	0 dB	0 dB
2,000 Hz		1 dB
4,000 Hz		1 dB
8,000 Hz	1 dB	

Definition **22B**:

The following Definition, **#22B**, identifies the specifics of the basic **Unitary Octave Band**, as specified by ANSI S1.11-1966 (R1975) — the nine **Unitary Octave Bands** listed on the previous page, namely, on Page 3-42, as part of the Definition of the **A-Weighted Sound Scale**, are the common **Unitary Octave Bands**. For this basic type of **Octave Band** (as well as every other category of **Octave Band**) the **Center Frequency** of the "middle" **Octave Band** is always 1,000 Hertz. The set of **Unitary Octave Bands** — or simply, the set of **Octave Bands** — are each characterized as: (1) - being one octave in total **Bandwidth**, ie. the **Lower Band-Edge Frequency** of each **Octave Band** will always be half of its **Upper Band-Edge Frequency**; and (2) - having a **Center Frequency** that will always be half of the **Center Frequency** of the next higher **Octave Band**, and twice the **Center Frequency** of the next lower one. Thus, the overall Definition is as follows:

$$f_{upper-1/1} = 2\,f_{lower-1/1}$$

$$f_{center-1/1} = \sqrt{f_{upper-1/1}\,f_{lower-1/1}} = \text{the "Geometric Mean"}$$

Where: $f_{upper-1/1}$ = the **Upper Band-Edge Frequency** for the specific **Unitary Octave Band** being considered;

$f_{lower-1/1}$ = the **Lower Band-Edge Frequency** for the specific **Unitary Octave Band** being considered; &

$f_{center-1/1}$ = the **Center Frequency** for the specific **Unitary Octave Band** being considered.

Definition **22C**:

The following Definition, **#22C**, identifies the specifics of the **Standard Half Octave Band**, also as specified by ANSI S1.11-1966 (R1975). The set of **Half Octave Bands** are each characterized as: (1) - being $1/\sqrt{2}$ Octaves in total Bandwidth, ie. the **Lower Band-Edge Frequency** of each **Half Octave Band** will always be $1/\sqrt{2}$ of its **Upper Band-Edge Frequency**; and (2) - having a **Center Frequency** that will always be $1/\sqrt{2}$ of the **Center Frequency** of the next higher **Half Octave Band**, and $\sqrt{2}$ times the **Center Frequency** of the next lower one. Thus, the overall Definition is as follows:

$$f_{upper-1/2} = \sqrt{2}\, f_{lower-1/2}$$

$$f_{center-1/2} = \sqrt{f_{upper-1/2}\, f_{lower-1/2}} = \text{the "Geometric Mean"}$$

Where: $f_{upper-1/2}$ = the **Upper Band-Edge Frequency** for the specific **Half Octave Band** being considered;

$f_{lower-1/2}$ = the **Lower Band-Edge Frequency** for the specific **Half Octave Band** being considered; &

$f_{center-1/2}$ = the **Center Frequency** for the specific **Half Octave Band** being considered.

<u>Definition **22D**</u>:

The following Definition, **#22D**, identifies, finally, the specifics of any "**1/n**th" **Octave Band**, again as specified by ANSI S1.11-1966 (R1975). The set of "**1/n**th" **Octave Bands** are each characterized as: (1) - being $1/\sqrt[n]{2}$ Octaves in <u>total</u> **Bandwidth**, ie. the **Lower Band-Edge Frequency** of each "**1/n**th" **Octave Band** will <u>always</u> be $1/\sqrt[n]{2}$ of its **Upper Band-Edge Frequency**; and (2) - having a **Center Frequency** that will <u>always</u> be $1/\sqrt[n]{2}$ of the **Center Frequency** of the next higher "**1/n**th" **Octave Band**, <u>and</u> $\sqrt[n]{2}$ times the **Center Frequency** of the next lower one. Thus, the overall Definition is as follows:

$$f_{upper-1/n} = \sqrt[n]{2}\, f_{lower-1/n}$$

$$f_{center-1/n} = \sqrt{f_{upper-1/n}\, f_{lower-1/n}} = \text{the "Geometric Mean"}$$

Where: $f_{upper-1/n}$ = the **Upper Band-Edge Frequency** for the specific "**1/n**th" **Octave Band** being considered;

                         $f_{lower-1/n}$ = the **Lower Band-Edge Frequency** for the specific "**1/n**th" **Octave Band** being considered; &

                         $f_{center-1/n}$ = the **Center Frequency** for the specific "**1/n**th" **Octave Band** being considered.

Ventilation Related:

Calculations Involving Gases Moving in Ducts:

Equation 23A:

This Equation, **#23A**, is one of the most basic relationships in the area of ventilation. It relates the **Volumetric Flow Rate** in a Duct to its **Cross Sectional Area**, and the gas **Velocity** (the **Duct Velocity**) in it.

$$Q = AV$$

Where:

Q = the **Volumetric Flow Rate** in the Duct under consideration, in cubic feet/minute (cfm);

A = the **Cross Sectional Area** of the Duct under consideration, in square feet (ft^2); &

V = the **Velocity**, or **Duct Velocity**, of the gases moving in the Duct, in feet/minute (fpm).

Equation 23B:

The following Equation, **#23B**, is the first of two basic relationships involving the **Velocity** of the gases that are flowing in a Duct; this one relates this parameter to the **Velocity Pressure** and the **Density** of the gases that are flowing in the Duct.

$$V = 1,096\sqrt{\frac{VP}{\rho}}$$

Where:

V = the **Velocity** of the gases moving in the Duct, or the **Duct Velocity**, measured in feet per minute (fpm);

VP = the **Velocity Pressure** of the gases moving in the Duct, measured in inches of water; &

ρ = the **Density** of the gases flowing in the Duct, measured in lbs per cubic foot (lbs/ft^3) — for reference, ρ for air = 0.075 lbs/ft^3.

Equation **23C**:

The following Equation, **#23C**, is the second of the two basic relationships involving the **Duct Velocity** of the gases that are flowing in a Duct; this one simply relating this parameter to the **Velocity Pressure** of the gases that are flowing in the Duct.

$$V = 4,005\sqrt{VP}$$

Where:
V = the **Velocity** of the gases moving in the duct, or the **Duct Velocity**, measured in feet per minute (fpm);

VP is the **Velocity Pressure**, and is as defined on Page 3-46.

Equations **23D, 23E, & 23F**:

The following three Equations, #'s **20D, 20E, & 20F**, are presented as a group because they constitute three **General Rules** that will <u>always</u> apply to the gases (usually air) that are flowing in a Duct.

Equation **23D**:

$$TP_1 = TP_2 + losses$$

Where:
TP_1 = the **Total Pressure** at Point #1 in the Duct, measured in inches of water;

TP_2 = the **Total Pressure** at Point #2 [downstream of Point #1] in the Duct, also measured in inches of water; &

$losses$ = **Pressure Losses** from various Duct internal factors (such as friction) that affect the flow of the gases in it, measured in inches of water. In general, these Loss Factors are usually known or provided in the form of a "Specific Pressure Loss or Drop per Unit Length of Duct", ie. as inches of water loss/lineal foot of duct.

Equation **23E**:

$$TP = SP + VP$$

Where:

TP is the **Total Pressure** in the Duct, and is exactly as defined, listed either as, TP_1 or TP_2, on Page 3-47;

SP = the **Static Pressure** in the duct, measured in inches of water; &

VP is the **Velocity Pressure** in the Duct, and is exactly as defined on Page 3-46;

Equation **23F**:

$$SP_1 + VP_1 = SP_2 + VP_2 + losses$$

Where:

SP_i is the **Static Pressure** at the **i**th point in the Duct, and is as defined directly above on this page — listed there simply as SP;

VP_i is the **Velocity Pressure** at the **i**th point in the Duct, and is as defined directly above on this page — listed there simply as VP; &

$losses$ are the **Pressure Losses** and are exactly as defined on Page 3-47.

Various Basic Calculations Involving Hoods:

<u>Equations **24A** & **24B**</u>:

The following two Equations, #'s **24A** & **24B**, are the relationships that provide the **Capture Velocity** (or the **Centerline Velocity**) of a square or a round faced Hood of a known **Cross Sectional Area**, either <u>without</u> Flanges [Equation **24A**], or <u>with</u> Flanges [Equation **24B**], at a specified **Distance** directly in front of the plane of the opening of the Hood being considered and evaluated.

Equation **24A** [Hoods WITHOUT Flanges]:

$$V = \frac{Q}{10x^2 + A}$$

Equation **24B** [Hoods WITH Flanges]:

$$V = \frac{Q}{0.75\left[10x^2 + A\right]} = \frac{4Q}{3\left[10x^2 + A\right]}$$

Where:

V = the **Capture Velocity** (or the **Centerline Velocity**) of the particular Hood under consideration, at a point "**X**" feet in front of the face of the Hood, this **Capture Velocity** is usually measured in feet per minute (fpm);

Q = the **Volumetric Flow Rate** of the Hood, measured in cubic feet per minute (cfm);

A = the **Cross Sectional Area** of the Hood opening, measured in square feet (ft^2); &

X = the **Distance**, measured in feet (ft), from the plane of the Hood opening to the point in front of the Hood where the **Capture Velocity** is to be determined.

Equation **24C**:

The following Equation, **#24C**, is usually referred to as the **Simple Hood Formula**. It is widely used to determine the **Hood Static Pressure**, which is one of the Prime Operating Factors that are applicable to any Hood.

$$SP_h = VP_d + h_e$$

Where:

SP_h = the **Hood Static Pressure**, measured in inches of water;

VP_d = the **Velocity Pressure in the Hood Duct**, also measured in inches of water; &

h_e = the **Hood Entry Loss Factor**, also measured in inches of water.

Equation **24D**:

The following Equation, **#24D**, is used to calculate the **Coefficient of Entry** for any Hood. This dimensionless parameter serves, functionally at least, as an Efficiency Rating for any Hood.

$$C_e = \sqrt{\frac{VP_d}{SP_h}} = \frac{\text{Actual Flow}}{\text{Theoretical Flow}}$$

Where:

C_e = the **Coefficient of Entry** for the Hood under investigation, this is a dimensionless parameter;

VP_d is the **Velocity Pressure in the Hood Duct**, and is as defined directly above on this page; &

SP_h is the **Hood Static Pressure**, and is also as defined directly above on this page.

<u>Equation **24E**</u>:

The following two forms of Equation, **#24E**, are known, collectively, as the **Hood Throat-Suction Equations**. They, too, are widely used to determine **Volumetric Flow Rates** of Hoods.

$$Q = 4,005A\sqrt{VP_d} = 4,005A\,C_e\sqrt{SP_h}$$

Where:

Q is the **Volumetric Flow Rate** and is as defined on Page 3-49;

A is the **Cross Sectional Area** of the Hood opening, and is also as defined on Page 3-49;

VP_d is the **Velocity Pressure in the Hood Duct**, and is as defined on Page 3-50;

C_e is the **Coefficient of Entry**, and is also as defined on Page 3-50; &

SP_h is the **Hood Static Pressure**, and is also as defined on Page 3-50.

Equation **24F**:

The following Equation, **#24F**, is known as the **Hood Entry Loss Equation,** and simply uses most of the previously defined parameters to develop a different and very useful relationship.

$$h_e = \left[\frac{1 - C_e^2}{C_e^2} \right] VP_h$$

Where:

h_e is the **Hood Entry Loss Factor,** and is as defined on **Page 3-50;**

C_e is the **Coefficient of Entry** for the Hood under consideration, and is also as defined on Page 3-50; &

VP_h = the **Velocity Pressure in the Hood,** measured in inches of water.

Equation **24G**:

The following Equation, **#24G**, is known as the **Hood Entry Loss Factor Equation,** and it, too, is widely used.

$$F_h = \frac{h_e}{VP_h}$$

Where:

F_h = the **Hood Entry Loss Factor,** which is a dimensionless parameter;

h_e is the **Hood Entry Loss Factor,** and is as defined on Page 3-50; &

VP_h is the **Velocity Pressure in the Hood,** and is as defined directly above on this page.

<u>Equation 24H</u>:

The following variations of Equation #24H are known as the **Compound Hood Equations**. They provide a vehicle for calculating the **Hood Static Pressure** for <u>all</u> conditions in which there are quantitative differences in the **Duct Velocity** and the **Hood Slot Velocity**. There are two circumstances that are considered below: <u>Condition #1</u>: the **Duct Velocity** is GREATER than the **Hood Slot Velocity**. <u>Condition #2</u>: the **Hood Slot Velocity** is GREATER than the **Duct Velocity**).

Equation **24H** [Condition #1]:

$$SP_h = h_{ES} + h_{ED} + VP_d \qquad \text{when:} \quad V_d > V_s$$

Equation **24I** [Condition #2]:

$$SP_h = h_{ES} + h_{ED} + VP_s \qquad \text{when:} \quad V_s > V_d$$

For most Hood and Duct situations, the following two approximations hold:

$$h_{ES} = 1.78[VP_s] \quad \& \quad h_{ED} = 0.25[VP_d]$$

Where:

SP_h is the **Hood Static Pressure**, and is as defined on Page 3-50;

h_{ES} = the **Hood Slot Entry Loss**, measured in inches of water;

h_{ED} = the **Hood Duct Entry Loss**, measured in inches of water;

VP_d = the **Duct Velocity Pressure**, measured in inches of water;

VP_s = the **Slot Velocity Pressure**, measured in inches of water;

V_s = the **Slot Velocity**, measured in velocity units, such as feet per minute (fpm); &

V_d = the **Duct Velocity**, measured in the same velocity units as V_s.

Calculations Involving the Rotational Speeds of Fans:

Equation 25A:

The following Equation, #25A, relates the **Air Discharge Volume** being provided by a Fan to its **Rotational Speed**. This is very important whenever one is trying to decide on the relative sizes of the pulleys involved (one on the Fan itself, the other on the motor that is to serve as the motive force for the Fan) that will be used. This relationship, as well as each of the five that will follow, assumes that the Fan being evaluated may well be used under **different** operating circumstances, but only while handling fluids or gases of the **same** density.

$$\frac{CFM_1}{CFM_2} = \frac{RPM_1}{RPM_2}$$

Where: CFM_i = the **Air Discharge Volume** of the Fan when it is operating at the ith set of conditions, measured in cubic feet per minute (cfm); &

RPM_i = the ith operating **Rotational Speed** of the Fan, in revolutions per minute (rpm).

Equation 25B:

The following Equation, #25B, relates the **Static Discharge Pressure** of a Fan to its **Rotational Speed**. As with the Equation above on this same page, this relationship assumes that the Fan will be evaluated only for gases or fluids of the same density.

$$\frac{SP_{FAN_1}}{SP_{FAN_2}} = \left[\frac{RPM_1}{RPM_2}\right]^2$$

Where: SP_{FAN_i} = the **Static Discharge Pressure** of the Fan when it is operating at the ith set of conditions; &

RPM_i is the ith operating **Rotational Speed** of the Fan, as defined above on this page.

Equation 25C:

The following Equation, **#25C**, relates the motive **Brake Horsepower** required to operate a Fan at any of its possible **Rotational Speeds**. As with the Equation on the previous page, this relationship assumes that the Fan will be evaluated only for different **Rotational Speed** applications that involve gases and fluids having the same density.

$$\frac{BHP_1}{BHP_2} = \left[\frac{RPM_1}{RPM_2}\right]^3$$

Where: BHP_i = the **Brake Horsepower** required for a Fan to be operated at the ith set of conditions, measured in brake horsepower units (bhp); &

RPM_i is the ith operating **Rotational Speed** of the Fan, as defined on Page 3-54.

Calculations Involving the Diameters of Fans:

<u>Equation **26A**</u>:

The following Equation, **#26A**, relates the **Air Discharge Volume** being produced by a Fan to its **Diameter**. This relationship, like the previous three, as well as each of the two that will follow, assumes that the Fan being examined will <u>only</u> be evaluated for **different** <u>Diameters</u>, while handling gases or fluids of the <u>**same**</u> <u>density</u>.

$$\frac{CFM_1}{CFM_2} = \left[\frac{D_1}{D_2}\right]^3$$

Where: CFM_i = the **Air Discharge Volume** of the Fan at its ith **Diameter**, this **Air Discharge Volume** measured in cubic feet per minute (cfm); &

D_i = the **Diameter** of the ith Fan, measured in inches.

<u>Equation **26B**</u>:

The following Equation, **#26B**, relates the **Static Discharge Pressure** being developed by a Fan to its **Diameter**. This relationship, like the previous four, assumes that the Fan being evaluated will <u>only</u> be operated at **different** <u>Diameters</u> while handling gases and fluids of the <u>**same**</u> density.

$$\frac{SP_{FAN_1}}{SP_{FAN_2}} = \left[\frac{D_1}{D_2}\right]^2$$

Where: SP_{FAN_i} = the **Static Discharge Pressure** of a Fan of the ith **Diameter**, this **Static Discharge Pressure** measured in inches of water; &

D_i = the **Diameter** of the ith Fan, as defined above on this same page.

Equation **26C**:

The following Equation, **#26C**, relates the **Brake Horsepower** required to operate Fans of various **Diameters**. This relationship — as was true with the previous
five — assumes that the Fan under evaluation will <u>only</u> be operated at **different**
<u>Diameters</u> while handling gases or fluids of the <u>same</u> <u>density</u>, and at the <u>same</u> **Fan
Discharge Volume**.

$$\frac{BHP_1}{BHP_2} = \left[\frac{D_1}{D_2} \right]^5$$

Where: BHP_i = the **Brake Horsepower** required for a Fan of
the **i**th **Diameter** to deliver some specified,
and for this application, identical **Air Discharge Volume**, measured in brake horsepower units (bhp); &

D_i the **Diameter** of the **i**th Fan, as defined on
Page 3-56.

Calculations Involving Various Other Fan Related Factors:

<u>Equation 27A</u>:

The following Equation, **#27A**, is commonly known as the **Fan Brake Horsepower Equation**. It combines most of the factors that must be considered in the choice of a ventilation Fan.

$$BHP = \frac{1}{63.56}\left[\frac{[CFM][TP]}{FME}\right] = 0.0157\left[\frac{[CFM][TP]}{FME}\right]$$

Where:

BHP = the **Brake Horsepower** that will be required for the Fan to be able to provide the required performance level, measured in brake horsepower units (bhp);

CFM = the required **Fan Discharge Volume**, measured in cubic feet per minute (cfm);

TP = the **Fan Total Pressure**, measured in inches of water; &

FME = the **Fan Mechanical Efficiency**, which is a dimensionless percentage $\leq 100\%$.

<u>Equation 27B</u>:

The following Equation, **#27B**, is known as the **Fan Total Pressure Equation**. It is a very simple, straightforward, and widely used relationship.

$$TP_{FAN} = TP_{out} - TP_{in}$$

Where:

TP_{FAN} = the **Fan Total Pressure**, measured in inches of water;

TP_{out} = the **Fan Total Output Pressure**, measured in inches of water; &

TP_{in} = the **Fan Total Input Pressure**, measured in inches of water.

Equation 27C:

The following Equation, **#27C**, is known as the **Fan Static Pressure Equation**, because it basically defines the overall **Fan Static Pressure**.

$$SP_{FAN} = SP_{out} - SP_{in} - VP_{in}$$

Where: SP_{FAN} = the **Fan Static Discharge Pressure**, measured in inches of water;

SP_{out} = the **Fan Outlet Static Pressure**, measured in inches of water;

SP_{in} = the **Fan Inlet Static Pressure**, measured in inches of water; &

VP_{in} = the **Fan Inlet Velocity Pressure**, also measured in inches of water.

Calculations Involving Air Flow Balancing at a Duct Junction:

<u>Equation 28A</u>:

The following family of Relationships and Equations, **#28A**, define the process of balancing a Duct Junction. This process is logical in its development and in the solution path it implies. In this procedure, consideration is given to the **Volumetric Flow Rate** in each of two Ducts that join to form a single larger Duct.

$$R = \frac{SP_{greater}}{SP_{lesser}}$$

Where:

R = the **Junction Balance Ratio**, which is a dimensionless number;

$SP_{greater}$ = the **Duct Static Pressure** in the duct that carries the *greater* Flow Volume into the Duct Junction, measured in inches of water; &

SP_{lesser} = the **Duct Static Pressure** in the Duct carrying the *lesser* Flow Volume into the Duct Junction, also measured in inches of water.

We must now consider <u>Three</u> Different Scenarios:

I. $R < 1.05$ The Duct Junction is considered to be balanced.

II. $1.05 \leq R \leq 1.20$ Increase the **Volumetric Flow Rate** in the *lesser* volume branch from Q_{former} to Q_{new}, according to the following:

$$Q_{new} = Q_{former} \sqrt{\frac{SP_{greater}}{SP_{lesser}}}$$

III. $R > 1.20$ The Duct Junction is so <u>imbalanced</u> that the Entire Ventilation System must be <u>completely redesigned</u> from the ground up.

Calculations Involving Dilution Ventilation:

Equation 29A:

The following Equation, #29A, provides a single relationship that makes it possible to calculate the steady state, **Equilibrium Concentration** that would be produced in a Room (or any enclosed space for which the Overall Volume can be determined) by the complete evaporation of some specific **Volume** of any identifiable Volatile Solvent.

$$C = \left[\frac{V_s \rho T}{[MW_s] P_{atm} V_{room}} \right] [6.24 \times 10^7]$$

Where:

C = the **Equilibrium Concentration** of the Volatile Solvent that would be produced in the Room by the evaporation of the known **Volume of Solvent**, C measured in ppm;

V_s = the **Volume of the Solvent** that has evaporated, measured in milliliters (ml);

ρ = the **Density** of the Solvent, measured in grams per cubic centimeter (gms/cm^3);

T = the **Temperature** in the Room, measured in degrees Kelvin (°K);

MW_s = the **Molecular Weight** of the Solvent;

P_{atm} = the **Ambient Barometric Pressure** that is prevailing in the Room, measured in millimeters of mercury (mmHg);

V_{room} = the **Volume of the Room**, measured in liters; &

6.24×10^7 = the **Proportionality Constant** that makes this Equation valid, under **NTP** conditions.

Equation **29B**:

The following Equation, **#29B**, is known as the **Basic Room Purge Equation**. It provides the necessary relationship for determining the **Time Required** to re- duce a known **Initial High Level Concentration** of <u>any</u> Vapor — existing in a defined Closed Space or Room — to a more acceptable **Ending Lower Level Concentration**.

$$
D_t = \left[\frac{V}{Q} \right] ln \left[\frac{C_{initial}}{C_{ending}} \right]
$$

Where:

D_t = the **Time Required** to reduce the Vapor con- centration in the Closed Space or Room, as required, measured in minutes;

V = the **Volume** of the Closed Space or Room, measured in cubic feet (ft^3);

Q = the **Ventilation Rate** at which the Closed Space or Room will be purged by whatever air handling system is available for that pur- pose, measured in cubic feet per minute (cfm);

$C_{initial}$ = the **Initial High Level Concentration** of the Vapor in the ambient air of the Closed Space or Room, which Concentration is to be reduced — by purging at Q cfm — to a more acceptable **Ending Lower Level Concentration**, measured in ppm; &

C_{ending} = the desired **Ending Lower Level Concen- tration** of the Vapor that is to result from the purging effort in the Closed Space or Room, also measured in ppm.

Equation **29C**:

The following Equation, **#29C**, is known as the **Purge - Dilution Equation**. It is the most basic and fundamental relationship available for determining the various parameters associated with reducing the Concentration of the Vapor of any volatile material in a Closed Space or Room.

$$C = C_0 e^{-\left[V_{removed}/V_{room}\right]}$$

Where:

C = the **Ending Concentration** of the Vapor in the Closed Space or Room, which **Ending Concentration**, measured in ppm, resulted from the purging activities;

C_0 = the **Initial Concentration** of the Vapor in the Closed Space or Room that is to be reduced by purging, also measured in ppm;

$V_{removed}$ = the **Air Volume** that has been withdrawn from the Closed Space or Room, measured in any suitable volumetric units, usually in cubic feet (ft^3); &

V_{room} = the **Volume of the Room**, measured in the same volumetric units as $V_{removed}$, which is usually in cubic feet (ft^3).

Statistics and Probability Related:

Parameters Relating to Normal Distributions:

Equation **30A**:

The following Equation, **#30A**, defines the first, and probably the most important and most widely used, measure of location – or Central Tendency – for <u>any</u> quantifiable parameter, for which the resultant Distribution can be considered to be Normal. This measure is called the **Mean** of that Distribution:

$$\mu = \overline{x} = \frac{1}{n} \sum_{i=1}^{n} x_i$$

Where: $\mu = \overline{x}$ = the **Mean** of the Distribution of "n" different values of x_i ;

     x_i = the **Value** of the ith of the "n" members of the overall Distribution; &

     n = the **Number** of Members in the Distribution.

Definition **30B**:

The following Definition, **#30B**, defines the second measure of location, or Central Tendency, for any quantifiable parameter, for which the resultant Distribution is Normal. This measure is called the **Median** of the Distribution:

m_e = the **Median** or "Midpoint" Value in a Normal Distribution of "n" different numeric Values of x_i , ie. that <u>specific</u> <u>numeric</u> <u>Value</u> of x_i for which there are as many Values <u>in the Distribution greater</u> than this number, as there are Values <u>in the Distribution less</u> than this number.

Where: m_e = the **Median** of the Distribution of "n" different values of x_i ;

     x_i is as defined above on this page; &

     n is also as defined above on this page.

Definition **30C**:

The following Definition, **#30C**, characterizes the third measure of location, or Central Tendency, for any quantifiable parameter, for which the resultant distribution can be considered to be Normal. This measure is called the **Mode** of that Distribution:

---

m_o = the **Mode** or "Most Populous" Value in a Normal Distribution of "n" different numeric Values of X_i, ie. that <u>specific</u> <u>numeric</u> <u>Value</u> of X_i which is the <u>most</u> <u>frequently</u> <u>occurring</u> Value in the entire Distribution.

Where: m_o = the **Mode** of the Distribution of "n" different Values of X_i ;

             X_i is as defined on Page 3-64; &

             n is also as defined on Page 3-64.

Equation **30D**:

The following Equation, **#30D**, defines the **Range** for any Data Set or Normal Distribution of quantifiable parameters. It is determined by subtracting the **Value of the Smallest Member** of the entire Data Set from the **Value of the Largest Member**.

$$R = \left[x_{i_{max}} - x_{i_{min}} \right]$$

Where: R = the **Range** of the Normal Distribution of "n" different Values of X_i ;

          $X_{i_{max}}$ = the **Value of the Largest Member** of X_i in the entire Distribution or Data Set; &

          $X_{i_{min}}$ = the **Value of the Smallest Member** of X_i in the entire Distribution or Data Set.

Equation 30E:

The following Equation, #30E, characterizes and defines the final measure of location — or Central Tendency — for any measurable or quantifiable parameter, for any Distribution (Normal or otherwise). This measure is called the **Geometric Mean** of that Distribution. It is a somewhat more useful measure than the simple **Mean**, when one must deal with any Distribution for which the Ratio of the **Value of the Largest Member** to the **Value of the Smallest Member** is greater than **200**.

$$M_{geometric} = \sqrt[n]{(x_1)(x_2)(x_3)\cdots(x_n)} = 10^{\frac{1}{n}\sum_{i=1}^{n} \log x_i}$$

for any Distribution, but most useful for those Distributions for which:

$$\frac{X_{i_{max}}}{X_{i_{min}}} \geq 200$$

Where: $M_{geometric}$ = the **Geometric Mean** of the Distribution under consideration;

X_i is as defined on Page 3-64;

n is also as defined on Page 3-64;

$X_{i_{max}}$ is the **Value of the Largest Member** of X_i and is as defined on Page 3-65; &

$X_{i_{min}}$ is the **Value of the Smallest Member** of X_i and is also as defined on Page 3-65.

<u>Equation 31A</u>:

The following identical set of Equations, **#31A**, defines the **Sample Variance**, which is the first measure of the Variability — or Dispersion — of <u>all</u> the data in <u>any</u> Normal Distribution.

$$ s^2 = \frac{\sum\limits_{i=1}^{n} \left[x_i - \mu \right]^2}{n-1} = \frac{\sum\limits_{i=1}^{n} \left[x_i - \overline{x} \right]^2}{n-1} $$

Where:

s^2 = the **Sample Variance** for the entire Data Set of "n" normally distributed values of x_i ;

x_i is as defined on Page 3-64;

n is also as defined on Page 3-64;

$\mu = \overline{x}$ the **Mean** of the Distribution or Data Set, and is also as defined on Page 3-64.

Equation **31B**:

The following identical set of Equations, **#31B**, defines the **Population Variance**, which is a second, and relatively less important, measure of the Variability — or Dispersion — of *all* the data in *any* Normal Distribution.

$$\sigma^2 = \frac{\sum_{i=1}^{n} \left[x_i - \mu \right]^2}{n} = \frac{\sum_{i=1}^{n} \left[x_i - \overline{x} \right]^2}{n}$$

Where: σ^2 = the **Population Variance** for the entire Data Set of "n" Normally Distributed Values of x_i;

x_i is as defined on Page 3-64;

μ = $\overline{x}$ the **Mean** of the Distribution or Data Set, and is also as defined on Page 3-64; &

n is also as defined on Page 3-64.

Equation **31C**:

The following set of identical Equations, **#31C**, defines the **Sample Standard Deviation**, which is the third, and probably the <u>most</u> <u>important</u>, measure of the Variability — or Dispersion — of <u>all</u> the data in <u>any</u> Normal Distribution.

$$s = \sqrt{s^2} = \sqrt{\frac{\sum\limits_{i=1}^{n}\left[x_i - \mu\right]^2}{n-1}} = \sqrt{\frac{\sum\limits_{i=1}^{n}\left[x_i - \overline{x}\right]^2}{n-1}}$$

Where:

s = the **Sample Standard Deviation** for the entire data set of "n" Normally Distributed Values of x_i;

s^2 is the **Sample Variance** and is as defined on Page 3-67;

x_i is as defined on Page 3-64;

$\mu = \overline{x}$ is the **Mean** and is also as defined on Page 3-64; &

n is also as defined on Page 3-64.

<u>Equation **31D**</u>:

The following set of identical Equations, **#31D**, defines the **Population Standard Deviation**, which is the fourth, and probably <u>least</u> <u>important</u>, measure of the Variability — or Dispersion — of <u>all</u> the data in <u>any</u> Normal Distribution.

$$\sigma = \sqrt{\sigma^2} = \sqrt{\frac{\sum_{i=1}^{n}\left[x_i - \mu\right]^2}{n}} = \sqrt{\frac{\sum_{i=1}^{n}\left[x_i - \overline{x}\right]^2}{n}}$$

Where:

σ = the **Population Standard Deviation** for the entire Data Set of "n" Normally Distributed Values of x_i ;

σ^2 is the **Population Variance**, and is as defined on Page 3-68;

x_i is as defined on Page 3-64;

$\mu = \overline{x}$ is the **Mean** and is also as defined on Page 3-64; &

n is also as defined on Page 3-64.

Equation **32A**:

The following set of identical Equations, **#32A**, defines the **Sample Coefficient of Variance**, which is the first measure of the Specific Dispersion of all the data in any Normally Distributed set of Values.

$$CV_{sample} = \frac{S}{\mu} = \frac{S}{\overline{X}}$$

Where: CV_{sample} = the **Sample Coefficient of Variance** for the Normally Distributed set of "n" different values of X_i ;

S is the **Sample Standard Deviation**, and is as defined on Page 3-69; &

$\mu = \overline{X}$ is the **Mean** and is as defined on Page 3-64.

Equation **32B**:

The following Equations, **#32B**, define the **Population Coefficient of Variance**, which is the second measure of the Specific Dispersion of all the data in any Normally Distributed set of Values. Proceeding logically from the previous Equation, this one is given by:

$$CV_{population} = \frac{\sigma}{\mu} = \frac{\sigma}{\overline{X}}$$

Where: $CV_{population}$ = the **Population Coefficient of Variance** for the Normally Distributed set of "n" different values of X_i ;

σ is the **Population Standard Deviation**, and is as defined on Page 3-70; &

$\mu = \overline{X}$ is the **Mean** and is as defined on Page 3-64.

Example Problems

Problem #1:

A worker was exposed, over his work shift, to various levels of n-pentane, and for varying periods of time, as shown in the tabulated exposure history that follows. The appropriate standards for n-pentane are: the TLV-TWA is 600 ppm, and the TLV-STEL is 750 ppm. What was his 8-hour TWA exposure to n-pentane?

2.0 hours	451 ppm
3.0 hours	728 ppm
1.5 hours	619 ppm
1.5 hours	501 ppm

Applicable Formula on Page 3-1   Problem Solution on Page 5-2

Problem #2:

During the concluding 15 minutes of the final 1.5 hour segment of his workday, what would this worker's exposure to n-pentane have to have been so that his overall 8-hour TWA would have been in violation of the established ACGIH TLV?

Work Space Continued on the Next Page

Continuation of Work Space for **Problem #2**

Applicable Formula on Page 3-1   Problem Solution on Page 5-3

Problem #3:

During the concluding 15 minutes of the final 1.5 hour segment of his workday, what would his exposure have to have been so that his work experience for that entire day would have involved a TLV-STEL violation? Would this 15 minute exposure also have produced a TLV-TWA violation?

Applicable Definitions on Pages 1-1 & 1-2   Problem Solution on Page 5-4

Problem #4:

An office machine repair technician was exposed to a relatively high concentration level of ozone from a malfunctioning copier he had been called upon to repair. He observed the operation of the machine for 30 minutes in order to diagnose its problems. He then fixed the problem by replacing a defective part. Finally, he observed the operation of the then <u>properly</u> <u>functioning</u> machine for an additional 30 minutes to ensure that it was, in fact, working properly. His ozone exposures for the day on which he fixed this machine were as follows:

<u>Task Description</u>	<u>Exposure Time</u>	<u>Ozone Exposures</u>
Diagnostic Effort	30 min	289 ppb
Repair Effort	60 min	42 ppb
Repair Observing Time	30 min	93 ppb
Balance of the Work Day	6 hours	8 ppb

The PEL-TWA for ozone is 0.1 ppm, and the PEL-STEL for this material is 0.3 ppm. What was this worker's 8-hour TWA to ozone? Did this worker experience a PEL-STEL violation?

Applicable Formula on Page 3-1   Problem Solution on Pages 5-4 & 5-5

Problem #5:

If the office machine repair technician from the previous problem worked in a geographic region that was characterized by a very bad "smog" condition, under which the exterior ambient air — on the particular day when this individual made the copier repairs described in **Problem #4** — had an ozone concentration of 222 ppb. If this technician spent 2.5 of the final 6 hours of that work day in his repair vehicle driving from one job to the next, and if all his other repairs on that day were to typewriters where he experienced only the previously mentioned 8 ppb ozone background level, what would his new 8-hour TWA to ozone have been? For reference, his timed ozone exposures were broken down as follows:

Task Description	Exposure Time	Ozone Exposures
Initial Copier Repairs from **Problem #4**	120 min	determine from **Problem #4**
Repair Effort on Typewriters	210 min	8 ppb
Driving Time between Jobs	150 min	222 ppb

Problem #6:

An Engineering Technician for a large metropolitan hospital spent one full 8-hour workday servicing the gas sterilizer the hospital's Central Supply Department. His ethylene oxide exposures were measured continuously by an Industrial Hygienist using an infrared spectrophotometer (the minimum detectable ethylene oxide level [MDL] for this analyzer is 0.4 ppm). Ethylene Oxide TWA measurements were made for various periods of time, and were reported by this IH as follows:

2.2 hours	0.7 ppm
1.4 hours	1.2 ppm
1.5 hours	1.6 ppm
2.9 hours	< 0.4 ppm

From the perspective that the 8-hour PEL-TWA for ethylene oxide is 1.0 ppm, and its 8-hour TWA Action Level is 0.5 ppm, would this employee's TWA exposure on the date in question have to be reported as a violation? What was his exposure?

Work Space Continued on the Next Page

Continuation of Work Space for **Problem #6**

Applicable Formula on Page 3-1   Problem Solution on Pages 5-6 & 5-7

Problem #7

The production employees in a custom fiberglass fabrications shop are routinely exposed to styrene vapors [for styrene, the PEL-TWA = 50 ppm, and the PEL-STEL = 100 ppm]. After one particularly busy day, an employee's styrene dosimeter indicated an 8-hour TWA styrene exposure of 48 ppm. If her 8-hour workday included two 5-minute coffee breaks and a 30 minute lunch period, all three of which were spent in a room where the ambient styrene concentration level never exceeded 0.1 ppm with the balance of her workday being spent in the production area, what must this employee's average styrene exposure have been during those time periods when she was actually involved in productive work?

Work Space Continued on the Next Page

4-6

Continuation of Work Space for **Problem #7**

Applicable Formula on Page 3-1   Problem Solution on Pages 5-7 thru 5-9

Problem #8:

The Safety Manager of a large Cold Storage Plant, designed for holding apples in a fresh condition over extended periods of time, determined that the workers in his company's cold rooms have been experiencing an overall 84% TWA exposure to carbon dioxide. If (1) each worker must spend 6 hours/day in these cold rooms; (2) if the PEL-TLV for CO_2 is 10,000 ppm; and (3) if the outside air, to which each worker is exposed for the remainder of the work day, contains an average of 425 ppm of CO_2, what must the average concentration of CO_2 be in the cold rooms?

Applicable Formula on Page 3-1   Problem Solution on Page 5-9

Problem #9:

If the Safety Manager of the foregoing Cold Storage Plant [**Problem #8**] has determined: (1) that his workers should <u>never</u> experience more than a 50% TLV exposure to CO_2, and (2) that the <u>minimum</u> CO_2 concentration he can permit in the cold rooms [in order to provide for <u>proper</u> apple storage], is 1.5% by volume, what then will be the <u>maximum</u> <u>time</u> <u>period</u> he will be able to allow his company's employees to work in the cold rooms each day from now on?

Applicable Formula on Page 3-1   Problem Solution on Pages 5-9 & 5-10

Problem #10:

John Smith, an Industrial Hygienist, working for the ABC Co. calculated the following TWA employee exposures, for the following list of four solvent vapors, for those workers who operate the company's paint spray booth:

No.	Solvent	TWA	TLV-TWA
1.	MIBK	12 ppm	50 ppm
2.	toluene	17 ppm	100 ppm
3.	methanol	55 ppm	200 ppm
4.	IPA	91 ppm	400 ppm

What was the %TLV exposure of these workers?

Applicable Formula on Page 3-2   Problem Solution on Pages 5-10 & 5-11

Problem #11:

Kraft Pulp Mill employees involved in bleaching kraft paper are potentially exposed to <u>both</u> chlorine <u>and</u> chlorine dioxide. For these two chemicals, the published TLV-TWAs and TLV-STELs are as follows:

<u>Bleaching Gas</u>	<u>TLV-TWA</u>	<u>TLV-STEL</u>
Chlorine	500 ppb	1,000 ppb
Chlorine Dioxide	100 ppb	300 ppb

On a heavy bleaching day, one employee's dosimeters indicated that his 8-hour TWA exposures to these two materials had been 0.42 ppm for chlorine, and 0.08 ppm for chlorine dioxide. What was this employee's overall %TLV exposure to these hazardous vapors? Was his employer in violation of any ACGIH Standards?

Work Space Continued on the Next Page

Continuation of Work Space for **Problem #11**

Applicable Formula on Page 3-2   Problem Solution on Pages 5-11 & 5-12

Problem #12:

A vapor degreasing solvent consists of the following four components in the following proportions:

Component	Weight % in Solvent	TLV-TWA	Molecular Weight
Freon 11	25 %	1000 ppm	137.38
Freon 113	55 %	1000 ppm	187.38
methyl chloroform	15 %	350 ppm	133.41
methylene chloride	5 %	50 ppm	84.93

Assuming that the composition of the degreasing vapor is equal to the composition of the liquid mixture, what is the Effective TLV (in mg/m^3 & ppm) in the vapor space for this vapor degreasing solvent mixture? You may assume that the vapor degreaser employing this mixture is operated at NTP. What would the ambient air concentrations for each of the four components of this mixture be (in mg/m^3 & ppm) if the calculated overall Effective TLV for the vapor space above this degreaser were actually to be achieved?

Work Space Continued on the Next Page

Continuation of Work Space for **Problem #12**

Work Space Continued on the next page

Continuation of Work Space for **Problem #12**

Applicable Formulae on Pages 3-3, 3-4, & 3-5   Problem Solution on Pages 5-12 thru 5-15

Problem #13:

A solution containing two volatile solvents is used widely in the wafer fabrications area of a large semiconductor manufacturer. This solution is made up, by weight, of 65% cellosolve [2-ethoxyethanol] and 35% t-butyl alcohol. Assuming that the composition of the vapor is equal to the composition of the liquid solvent mixture, what is the Effective TLV (in mg/m^3 & in ppm) for this mixture? You may also assume that the wafer fabrications area is operated at NTP. What will the ambient air concentrations for each of the two components of this mixture be (in mg/m^3 & ppm) when the __total__ vapor concentration in the wafer fabrications area is equal to this Effective TLV? The following data may be useful for you:

Component	Structural Formulae	TLV-TWA	Mol. Weight
cellosolve	$HO\text{-}CH_2\text{-}CH_2\text{-}O\text{-}CH_2\text{-}CH_3$	200 ppm	90.12
t-butyl alcohol	$[CH_3]_3\text{-}C\text{-}OH$	100 ppm	74.12

Work Space Continued on the Next Page

Continuation of Work Space for **Problem #13**

Work Space Continued on the Next Page

Continuation of Work Space for **Problem #13**

Problem #14:

A U. S. Navy shipboard refrigerant consists of a well defined mixture of the following three chlorofluorocarbons, made up according to the following proportions:

Chemical	Wt. % in Refrigerant	Molecular Weight	TLV-TWA
Freon 12	40 %	120.92	1,000 ppm
Freon 21	20 %	102.92	10 ppm
Freon 112	40 %	170.92	500 ppm

Obviously any leak of this pressurized liquid refrigerant will produce a vapor whose composition will be the same as that of the liquid material. What will be the Effective TLV for this mixture (in both mg/m^3 & ppm)? U. S. Navy ships operate their refrigeration spaces at NTP.

Work Space Continued on the Next Page

Continuation of Work Space for **Problem #14**

Work Space Continued on the Next Page

Continuation of Work Space for **Problem #14**

Applicable Formulae on Pages 3-3, 3-4, & 3-5   Problem Solution on Pages 5-17 thru 5-20

Problem #15:

The Navy uses a bulkhead mounted, digital readout, highly specific Freon 21 monitor (ie. it does not respond to any vapor other than Freon 21) in each of the shipboard refrigeration spaces that employ the special refrigerant described in **Problem #14**. The response range of each of these gas monitors is 0 - 200 ppm [actually, 0.0 to 199.9 ppm]. Their readout is in the following form "XXX.X". In the most important range — namely, 0 - 10.0 ppm of Freon 21 — its accuracy is ± 0.05 ppm. At what concentration must each of the bi-level audible alarm settings on this instrument be set? Typically, a bi-level Alert Alarm will be set at 60% of the effective TLV-TWA, and a second, higher level Evacuate Alarm, at 90% of this same effective TLV-TWA. As stated in **Problem #14**, Navy ships usually operate their refrigeration spaces at NTP.

Applicable Information from **Problem #14**   Problem Solution on Pages 5-20 & 5-21

Problem #16:

Glutaraldehyde is the "sterilant" used as the active ingredient in numerous proprietary Cold Sterilizing solutions that are, in turn, used by many hospitals and other health care facilities. The German MAK-TWA for this material is 0.8 mg/m^3. Express this mass based concentration in its volume based equivalent, assuming NTP. The molecular weight of glutaraldehyde is 100.12 amu.

Applicable Formula on Page 3-4   Problem Solution on Page 5-21

Problem #17:

The OSHA PEL-STEL for tetrahydrofuran is 735 mg/m^3. The PEL-TWA is 590 mg/m^3. Please express these concentrations in ppm(vol) units, assuming STP conditions. The formula weight of tetrahydrofuran is 72.11 amu.

Applicable Formula on Page 3-4   Problem Solution on Pages 5-21 & 5-22

Problem #18:

The 8-hour PEL-TWA and the Action Level for ethylene oxide are, respectively, 1.0 ppm and 0.5 ppm. Express these concentrations in mg/m^3, assuming NTP. The molecular weight of ethylene oxide is 44.05.

Applicable Formula on Page 3-5   Problem Solution on Page 5-22

Problem #19:

Benzene is a relatively unique volatile organic in that it has <u>three</u> published OSHA Permissible Exposure Limits — its PEL-TWA = 1.0 ppm, its PEL-STEL = 5.0 ppm, and its PEL-C = 25.0 ppm. Express these concentrations in mg/m^3, assuming STP conditions. Benzene's molecular weight is 78.11.

Applicable Formula on Page 3-5   Problem Solution on Pages 5-22 & 5-23

Problem #20:

The Safety Manager for a Building Materials Supply Company has determined that the TLV_{quartz} for respirable quartz fraction in the river gravel that is currently being crushed in his plant is 0.22 mg/m^3. What is the approximate percentage of respirable quartz in the crushed gravel?

Applicable Formula on Page 3-6   Problem Solution on Page 5-23

Problem #21:

What is the 8-hour T_{mix} for the mixed quartz species in a sample that contains 80% total silica, if the three categories of silica that are present in the sample are in the following ratio: [quartz] : [cristobalite] : [tridymite] = 18 : 13 : 9?

Applicable Formula on Page 3-7   Problem Solution on Page 5-24

Problem #22:

A 100 liter meteorological balloon is filled with helium at NTP. What will its volume become at an altitude of 50,000 ft where the barometric pressure is 175 millibars?

Applicable Formula on Page 3-8   Problem Solution on Page 5-25

Problem #23:

A spherical bladder has been filled with air at a pressure of one atmosphere. What must the external pressure on this bladder become in order to reduce its diameter by 75%.

Applicable Formula on Page 3-8   Problem Solution on Page 5-25

Problem #24:

A one liter balloon is filled to capacity at STP, and then placed in a commercial freezer which operates at -13°F. By how much will the volume of the balloon decrease?

Applicable Formula on Page 3-9   Problem Solution on Pages 5-26 & 5-27

Problem #25:

A soap bubble has a volume of 46 ml at a room temperature of 77°F. When this soap bubble passed through the beam of an IR heat lamp, the temperature of the air inside it increased. While passing through this beam, this soap bubble's volume increased to 46.5 ml, at which time it burst. By how much had the internal temperature of the air in this soap bubble increased at the moment it broke?

Applicable Formula on Page 3-9   Problem Solution on Page 5-27

Problem #26:

A capped soft drink bottle has an internal pressure of 950 mm Hg at a refrigerator temperature of 3°C. If this bottle is accidentally left, unprotected, in the trunk of a car during the summer, and undergoes a temperature increase to 45°C, what will its internal pressure be (in mm Hg)?

Applicable Formula on Page 3-9   Problem Solution on Pages 5-27 & 5-28

Problem #27:

In the morning, after a car has been parked overnight, its tires will typically be at the same temperature as the atmosphere. For one car, the morning ambient temperature was 56°F. On that particular morning, the pressure in its tires was 32 psig. What will this pressure become (in psig) after the car has been driven all day through a hot desert environment causing its tires to be heated to 175°F?

Applicable Formula on Page 3-9   Problem Solution on Pages 5-28 & 5-29

Problem #28:

A 1.0 liter cylinder containing chlorine at 5 atmospheres and 77°F is used to fill a balloon at STP. What will be the volume of the chlorine resistant balloon, after all the chlorine that was in the cylinder is in the balloon?

Applicable Formula on Page 3-10   Problem Solution on Page 5-29

Problem #29:

A partially filled cylinder, having an internal volume of 30 liters, contained N_2O, and was located in the sunlight. It showed an internal pressure of 450 psia. This cylinder was used to provide a flow of laughing gas — for anesthesia purposes — at a flow rate of 3,000 cc/min for a time interval of 4 hours and 30 minutes, or until its internal pressure had dropped to 766 mm Hg. If this flow was delivered at a pressure of 766 mm Hg, and a temperature of 20°C, what must have been the temperature of the sun heated cylinder at the start of the process?

Applicable Formula on Page 3-10   Problem Solution on Pages 5-29 & 5-30

Problem #30:

Approximately 34 grams of a refrigerant occupy 8.58 liters at a temperature of 18°C and a pressure of 0.92 atmospheres. It is known that this refrigerant is one of the following three. Please identify the correct material and justify your choice.

No.	Refrigerants	Formula	Molecular Weight
1.	Freon 12	CCl_2F_2	120.91
2.	Freon 21	$CHFCl_2$	102.92
3.	Freon 22	$CHClF_2$	86.48

Applicable Formula on Page 3-11   Problem Solution on Pages 5-30 & 5-31

Problem #31:

What is the "Effective Molecular Weight" of the Air if 4.36 liters of it weigh 5.00 grams at a temperature of 31°C and a pressure of 755 mm Hg?

Applicable Formula on Page 3-11   Problem Solution on Pages 5-31 & 5-32

Problem #32:

What will be the weight of 4.4 moles of Chile Salt Peter, NaNO3? The following atomic weights may be of interest:

Element	Atomic Weight
Sodium	24.31 amu
Nitrogen	14.01 amu
Oxygen	16.00 amu

Applicable Formula on Page 3-12   Problem Solution on Page 5-32

Problem #33:

The actual weight of an individual atom of the most common isotope of copper is 1.045×10^{-13} nanograms [1 gram = 1,000,000,000 nanograms]. What is the atomic weight of this isotope of copper?

Applicable Formulae on Pages 3-12 & 3-13   Problem Solution on Pages 5-32 & 5-33

Problem #34:

The STP volume of a sample of helium gas [atomic weight = 4.00 amu] was found to be 3,400 cc. How many moles of helium were in the sample, and what did it weigh?

Applicable Formulae on Pages 3-12 & 3-14   Problem Solution on Page 5-33

Problem #35:

One of the most commonly used hospital gas sterilants is a product known as Oxy-fume 12®. This material is sold in pressurized cylinders and is blended by Union Carbide, its manufacturer, to contain 12%, by weight, ethylene oxide [Mol. Wt. = 44.05] and the balance, Freon 12 [Mol. Wt. = 120.92]. What are the proportions of this gas mixture when expressed in % by volume?

® Registered Trademark of the Union Carbide Corporation

Work Space Continued on the Next Page

Continuation of Work Space for **Problem #35**

Applicable Formula on Page 3-12   Problem Solution on Page 5-34

Problem #36:

In order to make Phineas Fogg's hot air balloon lift its 1,650 lbs load, it is necessary to decrease the density of the air in the balloon envelope by only 5%. If Mr. Fogg wishes to start his "80-Days Around-The-World" odyssey on a day when the ambient air temperature is 55°F, by how many degrees Fahrenheit will he have to increase the temperature of the gas in the balloon envelope to achieve a lift off?

Applicable Formula on Page 3-15   Problem Solution on Pages 5-35 & 5-36

Problem #37:

The nominal density of air under NTP conditions is 1.182 mg/cm^3. What will its density be at the booster inlet in a turbocharged internal combustion gasoline engine, where the temperature is 1,400°F and the pressure is 1,400 mm Hg?

Applicable Formula on Page 3-15   Problem Solution on Page 5-36

Problem #38:

Bone dry air contains three principal components. It is made up of 78.1% nitrogen, 20.9% oxygen, and 0.9% argon. The remaining balance of 0.1% is comprised of carbon dioxide, neon, and methane. Expressed in mm Hg, what is the partial pressure of oxygen at NTP? of argon?

Applicable Formulae on Pages 3-17   Problem Solution on Pages 5-36 & 5-37

Problem #39:

The Lower Explosive Limit (LEL) for propane in air is 2.1%. Its Upper Explosive Limit (UEL) is 9.5%. Expressing your answer in millibars, what would be the STP partial pressures of propane be at each of these concentration levels?

Applicable Formula on Page 3-17   Problem Solution on Page 5-37

Problem #40:

The vapor pressure of pure ethanol at 20°C is 44 mm Hg. 90 Proof Irish Whiskey contains 45% ethanol and 55% water, by weight. What would be the partial vapor pressure of ethanol above a shot glass that contains 200 ml of 90 Proof Irish Whiskey in a closed room at a temperature of 20°C, and a barometric pressure of 765 mm Hg? If this shot glass of Irish Whiskey were left uncovered in this closed room until it had come fully to evaporative equilibrium, what would the ambient concentration of ethanol be, in ppm? Should you have any use for the molecular weights of ethanol and water, they are: $MW_{ethanol} = 46.07$ & $MW_{water} = 18.02$.

Work Space Continued on the Next Page

Continuation of Work Space for **Problem #40**

Applicable Formulae on Pages 3-17 & 3-18   Problem Solution on Pages 5-37 & 5-38

Problem #41:

It is widely recognized that the specific gravity of Irish Whiskey is the same as that of water. It is desired to add distilled water to the shot glass listed in **Problem #40** in order to reduce the ultimate ambient equilibrium concentration level of ethanol that could be achieved in the air of the closed room. If the target ambient concentration level of ethanol is to be its PEL-TWA of 1,000 ppm, how much distilled water must be added to the shot glass? *Note: It is <u>highly</u> <u>unlikely</u> that <u>any</u> Irishman would enjoy such a dilution of his favorite beverage!*

Continuation of Work Space for **Problem #41**

Applicable Formulae on Pages 3-17 & 3-18   Problem Solution on Pages 5-39 & 5-40

Problem #42:

The owner of a metallic spring manufacturing company was advised by an Industrial Hygienist that the beryllium dust concentration on his stamping floor was 0.0032 mg/m^3, which was 160% of the PEL-TLV of 0.002 mg/m^3 for this material. This IH determined: (1) that the density of beryllium particulates was 1.848 gms/cm^3; and (2) that the average diameter of the particles in question in this factory was 1.5 microns. What settling velocity did this Industrial Hygienist determine for the beryllium particulates in this facility?

Applicable Formula on Page 3-19   Problem Solution on Pages 5-40 & 5-41

Problem #43:

The Industrial Hygienist from **Problem #42**, has further determined that the principal source of the beryllium dust is the facility's stamping presses. He determined that coatings of a silicone agent on the alloys to be stamped, would minimize the dust generation, primarily by causing small dust particles to agglomerate into significantly larger ones. If this Industrial Hygienist has determined that he must recommend a procedure that will increase the settling velocity of the beryllium dust particles to a <u>minimum</u> of 72 inches/hour, which of the following choices will make the greatest economic sense in solving this problem?

<u>Silicone Designation</u>	<u>Guaranteed Particle Diam. Increase/mg applied</u>	<u>Cost/gram</u>
A	1.90 X	$12.50
B	2.00 X	17.00
C	2.10 X	24.00

Applicable Formula on Page 3-19   Problem Solution on Page 5-41

Problem #44:

The mid-infrared wavelength at which the carbon-hydrogen bond absorbs (ie. the "carbon-hydrogen stretch") is at approximately 3.35 microns. What is the frequency, in hertz, of this wavelength?

Applicable Formula on Page 3-20   Problem Solution on Page 5-42

Problem #45:

One of the two γ-ray photons emitted during the decay of $_{27}Co^{60}$ has frequency of 2.84×10^{14} MHz. What is its wavelength, in microns?

Applicable Formula on Page 3-20   Problem Solution on Page 5-42

Problem #46:

What are the wavenumbers, in cm^{-1}, of the two photons identified in **Problem #**'s **44 & 45?**

Applicable Formula on Page 3-21   Problem Solution on Pages 5-42 & 5-43

Problem #47:

The radioactive isotope, $_{53}I^{131}$, is frequently used in the treatment of thyroid can-cer. It has a decay constant of 0.0862 days^{-1}. A local hospital has received its order of 2.0 µg of this isotope on January 1st. How much of this isotope will remain on January 20th of the same year? How much will remain on the one year anniversary [not a leap anniversary] of the receipt of these 2.0 µg of $_{53}I^{131}$?

Applicable Formula on Page 3-22   Problem Solution on Page 5-43

Problem #48:

The smallest amount of $_{99}Es^{245}$ that can be detected, or used in certain types of experimentation, is 1.5×10^{-10} ng. The radioactive decay constant for this isotope is 0.502 minutes^{-1}. If 8.8×10^{-6} ng of this material was successfully accumulated by a research scientist, how much time will this scientist have available to her as she performs experiments with her supply of this isotope — before it has decayed to the "barely detectable" level?

Applicable Formula on Page 3-22   Problem Solution on Page 5-44

Problem #49:

The heaviest hydrogen isotope, tritium, $_1H^3$, is radioactive. Its Half Life is 12.26 years. What is its radioactive decay constant?

Applicable Formula on Page 3-23   Problem Solution on Page 5-44

Problem #50:

What are the Half Lives of the two isotopes, $_{53}I^{131}$ and $_{99}Es^{245}$, that were identified in **Problem #'s 47 & 48**?

Applicable Formula on Page 3-23   Problem Solution on Pages 5-44 & 5-45

Problem #51:

$_{95}Am^{241}$, is one of the most commonly used, and readily available radioactive isotopes. It is, for example, widely used in commercial smoke detectors. Its Half Life is 432.2 years. A functional smoke detector must have <u>at least</u> 1.75 μg of this isotope in order to operate properly. If there are 4.0 x 10^{15} Americium atoms in each microgram of this isotope, what is the minimum number of disintegrations per second that are required to operate a smoke detector?

Work Space Continued on the Next Page

Continuation of Work Space for **Problem #51**

Applicable Formula on Page 3-24   Problem Solution on Pages 5-45 & 5-46

Problem #52:

If each smoke detector starts out with 1.80 µg of $_{95}Am^{241}$, how long will it be before the minimum required level of radioactive disintegrations per second has been reached? Has the manufacturer successfully built in product obsolescence?

Work Space Continued on the Next Page

Continuation of Work Space for **Problem #52**

Applicable Formulae on Pages 3-22 & 3-23   Problem Solution on Pages 5-46 & 5-47

Problem #53:

What is the Dose Exposure Rate, in mRads/day, that would be produced on a target that is 1.0 meter distant from a 550 mCi $_{11}Na^{24}$ source? The Radiation Constant for $_{11}Na^{24}$ is 18.81 Rad·cm^2/hr·mCi.

Applicable Formula on Page 3-25   Problem Solution on Page 5-47

Problem #54:

A Dose Exposure Rate of 48.0 mRads/hour was determined for a 440 µCi source of $_{88}Ra^{226}$ at a distance of 300 mm. What is the Radiation Constant for $_{88}Ra^{226}$, in units of Rad·cm²/hr·mCi?

Applicable Formula on Page 3-25   Problem Solution on Pages 5-47 & 5-48

Problem #55:

$_{27}Co^{60}$ decays by emitting γ-rays. It is widely used as a radiation source in the treatment certain cancerous tumors. The Radiation Technician who operates the Cobalt Radiation Source Tumor Treatment Apparatus [the CRSTTA] at a major hospital accumulates a steady 0.09 mRad/hr, as an absorbed dose, for each hour he is in the room with the CRSTTA when its aperture is closed (ie. when no patient treatment is occurring). This Technician's absorbed dose increases to 0.44 mRad/hr whenever the CRSTTA's aperture is opened and a patient is actually being treated. A typical day will have 2.2 hours of open aperture treatment time, and 5.8 hours of closed aperture operations (ie. set-up, dosimeter development, etc.). What would be this Technician's adjusted radiation dose, expressed in rem, during such an average day?

Work Space Continued on the Next Page

Continuation of Work Space for **Problem #55**

Applicable Formula on Page 3-26   Problem Solution on Page 5-48

Problem #56:

A large research facility has — among the numerous pieces of equipment that it makes available to its staff of Research Scientists for their various projects — a pool type nuclear reactor, equipped with an externally accessible graphite "Thermal Column".   The Technicians who work 8-hours each day, 5 days each week, in this area accumulate a background level absorbed radiation dose [of thermal neutrons] at a rate of 0.12 grays/hour. Whenever the access port to this Thermal Column is opened — in order to work on one of the experiments in it — each Technician's absorbed dose rate increases to 0.83 grays/hour. If the access port to this unit is opened for only 2 hours per week, with it remaining closed for the balance of the time, what will be the adjusted radiation dose rate that will be accumulated by each Technician who works in this area, expressed in sieverts/week?

Work Space Continued on the Next Page

Continuation of Work Space for **Problem #56**

Applicable Formula on Page 3-26

Problem Solution on Pages 5-48 & 5-49

Problem #57:

The facility in which the Technician from **Problem #55** works was equipped with an improved shielded cell in which he could operate the CRSTTA when its aperture was open. As a result, his radiation dose was reduced. The previous cell had been made from concrete, and was 18 inches thick. The improved cell was made from lead, and was 8 inches thick. If: (1) the Half Value Thickness for concrete is 2.45 inches; (2) the radiation emission rate for the $_{27}Co^{60}$ Source in the CRSTTA is 40 Rads/hour; (3) the Technician - Source geometry is completely unchanged [except for the new improved shielding cell]; and (4) the observed radiation emission rate for this facility has been reduced by a factor of 775 in the new cell, what is the Half Value Thickness of lead when used to shield the γ-rays from a $_{27}Co^{60}$ source?

Applicable Formula on Page 3-27   Problem Solution on Pages 5-49 & 5-50

Problem #58:

A patient who has been administered a quantity of $_{53}I^{131}$ for treatment of a thyroid cancer problem will, himself, become a radiation source for the γ-radiation emanating from this isotope as it accumulates in his or her cancerous thyroid gland. A doctor examining this patient from a distance of 15 cm would experience γ-radiation at an intensity of 200 Sieverts/hour. If this doctor were able to complete his examination at a distance of 25 cm, what would be the intensity of the radiation he would be experiencing at this increased distance?

Applicable Formula on Page 3-28   Problem Solution on Pages 5-50 & 5-51

Problem #59:

The adjusted radiation dose of a commercially available product that contains a low level $_{88}Ra^{266}$ source [an Alpha emitter] is 40 mRem/hour at a distance of 25 cm. If the radioactive source in this product is changed to $_{19}K^{45}$ [a β-emitter] of the same low level, what will be the effect on the adjusted radiation dose that would be experienced at the same 25 cm distance? What would be the distance from the altered product, containing the new potassium source, at which the adjusted radiation dose would be equal to the 40 mRem/hour level that was the case for the radium source?

Applicable Formulae on Pages 3-26 & 3-28   Problem Solution on Pages 5-51 & 5-52

Problem #60:

The Operator of a laser based industrial metal trimmer wears goggles that reduce the incident laser beam intensity at his eyes from 475 mW/cm^2 [at which level, he would not be able even to see the area where he was trimming] to a more workable and safe level of 0.45 mW/cm^2, as the transmitted intensity. What is the effective Optical Density of the protective goggles?

Applicable Formula on Page 3-29   Problem Solution on Page 5-52

Problem #61:

If the industrial metal trimmer identified in **Problem #60**, above, were to be retrofitted with a new improved laser source that had 6.62 times the laser beam intensity of the original, and if it is hoped to reduce still further the transmitted beam intensity to a maximum of 0.19 mW/cm^2, by how much must the Optical Density of the goggles be increased to satisfy these new conditions and requirements?

Applicable Formula on Page 3-29   Problem Solution on Pages 5-52 & 5-53

Problem #62:

The UHF Microwave Systems that are used for transmitting a very large fraction of all the public telecommunications in the United States employ a circular antenna with a diameter of 40.5 inches. If these antennas transmit their information using UHF Microwaves that have a wavelength of 46 cm, what will be the distance from these antennas to the Far Field?

Applicable Formula on Page 3-30

Problem Solution on Page 5-53

Problem #63:

If the average power output of the highly focused, very directional UHF Antennas from **Problem #62** above, is 0.05 kilowatts, what will be the approximate Power Density produced by this antenna at a point 12 inches directly in front of it? If the 6-minute TLV-TWA for this frequency [f ~ 600 MHz] is 6 mw/cm^2, for what maximum period can a Service Technician safely work 12 inches in front of this antenna when it is transmitting?

Work Space Continued on the Next Page

Continuation of Work Space for **Problem #63**

Applicable Formula on Page 3-31   Problem Solution on Pages 5-53 thru 5-55

Problem #64:

In order never to exceed the aforementioned TLV-TWA, what is the closest distance (directly in front of one of these UHF Transmitting Antennas) that a Service Technician may safely work?

Applicable Formula on Page 3-32   Problem Solution on Page 5-55

Problem #65:

If these "line-of-sight" UHF Microwave Antenna Systems can successfully transmit voice or digital data over a distance 85 miles, what is the minimum Power Density Level at which the system's receiving antenna can still be expected to operate successfully?

Applicable Formula on Page 3-32   Problem Solution on Pages 5-55 & 5-56

Problem #66:

Highway Patrol Officers frequently use an X-Band Radar system to measure the speed of vehicles on the highway. If their X-Band Speed Radar Guns operate at a microwave frequency of 10.525 GHz, and if they radiate from a 4.02 inch diameter antenna, how far will it be to the Far Field for such a Speed Radar Gun?

Applicable Formula on Page 3-30   Problem Solution on Pages 5-56 & 5-57

Problem #67:

If the Speed Radar Gun from **Problem #66** has an output power of 45 mw, at what minimum distance in front of this Gun's antenna must one be to ensure that he will never be in an area where the 6-minute TLV-TWA of 10 mw/cm^2 can be exceeded?

Applicable Formula on Page 3-31   Problem Solution on Page 5-57

Problem #68:

What would be the Wet Bulb Globe Temperature Index, in °C, for a Quarry Worker in Connecticut, who must work on a sunny summer morning when the Outdoor Dry Bulb Temperature is 88°F; the Wet Bulb Temperature, 72°F; and the Globe Temperature, 102°F?

Applicable Formula on Page 3-34   Problem Solution on Page 5-58

Problem #69:

Later in the same afternoon, in the same quarry identified in **Problem #68**, rain clouds have gathered, and rain has commenced to fall. The Quarry Manager has covered the work area in the pit with a large tarpaulin, to protect his employees. If the Wet Bulb Temperature under the tarp has increased to 78°C, while the Globe Temperature has remained unchanged. What will be the new WBGT Index for this slightly different situation?

Applicable Formula on Page 3-33 Problem Solution on Page 5-58

Problem #70:

In a large midwest Steel Mill, the corporate Industrial Hygienist has undertaken to evaluate the Heat Stress conditions to which each of the Mill's Open Hearth Operators are routinely exposed. These fully acclimatized workers are required, by their job description, to spend certain time periods in areas of the facility that are, quite understandably, <u>very hot</u>. For each time period spent in an area of great heat stress, these operators are required to spend a compensating time period in a cool rest area. This IH has determined site specific temperature conditions in this mill, as follows:

Location	Wet-Bulb Temperature	Globe Temperature
Open Hearth Area	117° F	175° F
Elsewhere in the Mill	102° F	93° F
Operator Rest Area	63° F	75° F

A typical Open Hearth Operator spends six 12-minute periods of each workday in the Open Hearth Area, and six corresponding 35-minute periods in the Rest Area. If these operators spend the balance of each 8-hour workday elsewhere in the Mill, what did the IH determine for the Time Weighted Average WBGT Index for a typical Open Hearth Operator?

Work Space Continued on the Next Page

Continuation of Work Space for **Problem #70**

Applicable Formulae on Pages 3-33 & 3-35   Problem Solution on Pages 5-58 & 5-59

Problem #71:

The Industrial Hygienist in **Problem #70** from the previous page recommended that the Time Weighted Average WBGT Index for a typical Open Hearth Operator be reduced. His recommendation was that the Steel Mill hire an additional Open Hearth Operator for each Shift. Doing so would decrease from six to five the number of time periods that each of these operators would have to spend in the Open Hearth Area. This IH also recommended that each compensating rest period be increased from 35-minutes to 45-minutes, recognizing that the balance of the time for each of these operators would still be spent elsewhere in the Mill. If these recommendations were implemented, what would the new improved Time Weighted Average WBGT Index be for a typical Open Hearth Operator?

Applicable Formula on Page 3-35   Problem Solution on Pages 5-59 & 5-60

Problem #72:

The average noontime, unshaded summer temperature in the Mojave Desert is 129°F. What is the approximate speed of sound in the Mojave Desert under these conditions?

Applicable Formula on Page 3-36   Problem Solution on Page 5-60

Problem #73:

On a calm day in January of any year, in Fairbanks, AK, the noontime temperature will typically be -35° C. What will be the speed of sound in air under such conditions?

Applicable Formula on Page 3-36   Problem Solution on Pages 5-60 & 5-61

Problem #74:

The maximum continuous noise level that is permitted by OSHA is 115 dB (at this level, the maximum permitted duration of this sort of continuous noise is limited to 7.5 minutes). What is the Analog Sound Pressure Level, in nt/m^2, of noise at this loudness level?

Applicable Formula on Page 3-36   Problem Solution on Page 61

Problem #75:

The average Analog Sound Intensity Level of a hummingbird hovering has been measured to be 2.45×10^{-7} watts/cm^2 at a distance of 2.0 meters. What is the corresponding Sound Intensity Level, in dB, at this distance?

Applicable Formula on Page 3-37   Problem Solution on Pages 5-61 & 5-62

Problem #76:

The Sound Power Level of a top fuel dragster (at maximum engine and super-charger RPM) is 134 dB. To what Analog Sound Power Level, expressed in watts, does this measured Sound Power Level correspond?

Applicable Formula on Page 3-37 · Problem Solution on Page 5-62

Problem #77:

At a distance of 300 feet, what sound pressure level, in dB, would a ground observer, without hearing protection, experience if he were to witness and listen to the takeoff of a US Navy F8U single jet engine fighter-interceptor? At takeoff, the sound pressure level of this aircraft's afterburner assisted jet engine, which can be regarded as being directly on the ground, is 165 dB.

Applicable Formula on Page 3-38 · Problem Solution on Pages 5-62 & 5-63

Problem #78:

At what altitude, measured in feet (directly above the observer), would the fighter listed in **Problem #77** have to pass, in order for the observer to experience the identical sound pressure level that was calculated for the previous problem, from the F8U's afterburner assisted jet engine?

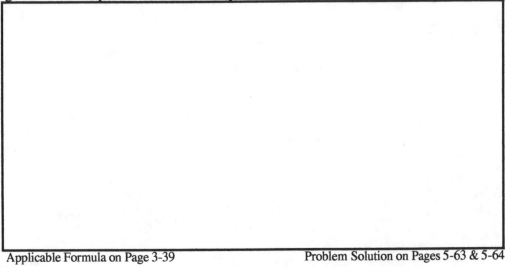

Applicable Formula on Page 3-38   Problem Solution on Page 5-63

Problem #79:

The Foreman of a Machine Shop has his work station located an equal distance from six separate grinders, each of which produces noise at 106 dB when in operation. What Sound Pressure Level, in dB, would the Foreman experience if all six grinders were operated simultaneously?

Applicable Formula on Page 3-39   Problem Solution on Pages 5-63 & 5-64

Problem #80:

The Director of a 20 member bagpipe band experiences a total Sound Pressure Level of 109 dB when he directs his ensemble. Assuming that each bagpipe produces music (??) at the same sound level as every other one, what must be the Sound Pressure Level of each instrument?

Applicable Formula on Page 3-39 Problem Solution on Pages 5-64 & 5-65

Problem #81:

How much longer is an individual, without hearing protection, permitted to work at a location where the noise level has just been reduced from 104 dB to 92 dB?

Applicable Formula on Page 3-40 Problem Solution on Page 5-65

Problem #82:

Standard ear plugs can reduce the sound of a band saw by 24 dB. Ear muffs can reduce the sound of this saw by 31 dB. A Band Saw Operator wearing ear plugs can safely operate her band saw for 4.6 hours per day. If she changes to using ear muffs, for how long a period will she be able to operate her band saw?

Applicable Formula on Page 3-40   Problem Solution on Pages 5-65 thru 5-67

Problem #83:

What is the Daily Dose, expressed as a percentage, for a worker who operates a lathe for 1.5 hours per day, sets his lathe up for 4.5 hours per day, performs administrative tasks for 1 hour per day, and spends the balance of his 8-hour workday either at breaks or eating his lunch? The average noise levels given below were determined by a competent Industrial Hygienist:

Task	Average Sound Pressure Level
Lathe Operation:	95 dB
Lathe Set-up:	90 dB
Breaks, Lunch, etc.	84 dB
Administrative Tasks	82 dB

Applicable Formulae on Pages 3-40 & 3-41   Problem Solution on Pages 5-67 & 5-68

Problem #84:

What is the Equivalent 8-hour Sound Pressure Level experienced by the Lathe Operator in **Problem #83**, above?

Applicable Formula on Page 3-42 Problem Solution on Page 5-68

Problem #85:

Four Printers work on a printing production floor where there are three offset presses. The Sound Pressure Levels, as a function of the number of these presses that are in operation, were determined to be as follows:

Number of Presses Operating	Average Sound Pressure Level	Average Daily Time in Operation
0	81 dB	4.5 hrs
1	93 dB	2.1 hrs
2	96 dB	1.0 hrs
3	98 dB	0.4 hrs

What is the Daily Dose that these Printers are experiencing? Is their Printing Company employer in violation of any OSHA Sound Pressure Level PEL?

Work Space Continued on the Next Page

Continuation of Work Space for **Problem #85**

Problem #86:

What is the Equivalent 8-hour Sound Pressure Level experienced by the three Printers listed above in **Problem #85**?

Applicable Formula on Page 3-42   Problem Solution on Page 5-70

Problem #87:

A monochromatic tuning fork, operating at "C-below-Middle-C" [$f_{lower\ C} = 261$ Hz], is observed to produce this tone at an Analog Sound Pressure Level of 71 dB, measured on the linear scale. What would a well calibrated Sound Level Meter, operating on the A-Scale, indicate as the Sound Pressure Level of this tuning fork?

Applicable Definitive Description on Page 3-42   Problem Solution on Pages 5-70 & 5-71

Problem #88:

What are the Upper and Lower Band Edge Frequencies of the only Octave band on the A-Scale that does not have a Sound Pressure Level adjustment?

Applicable Definitive Descriptions on Pages 3-42 & 3-43   Problem Solution on Page 5-71

Problem #89:

What is the Center Frequency of the Standard Unitary Octave Band, for which the Lower Band-Edge Frequency is 2,828 kHz? Justify your choice quantitatively.

Applicable Definitive Description on Page 3-43   Problem Solution on Pages 5-71 & 5-72

Problem #90:

What are the Upper and Lower Band-Edge Frequencies of the Standard One Half Octave Band that has a Center Frequency of 354 Hz?

Applicable Definitive Description on Page 3-44   Problem Solution on Page 5-72

Problem #91:

What is the Center Frequency of the Standard One Third Octave Band for which the Lower Band-Edge Frequency is 1,122 Hz?

Applicable Definitive Description on Page 3-45   Problem Solution on Page 5-73

Problem #92:

What is the Volumetric Flow Rate, in cfm, that exists in a 10 inch diameter duct in which air has been determined to be flowing at a velocity of 2,500 fpm?

Applicable Formula on Page 3-46   Problem Solution on Pages 5-73 & 5-74

Problem #93:

The main supply duct in the HVAC System in a new 40-story office building has Volumetric Flow Rate of 25,135 cfm. If this duct has air flowing in it at a velocity of 8,000 fpm, what is its diameter?

Applicable Formula on Page 3-46   Problem Solution on Page 5-74

Problem #94:

Assuming that the Density of the air flowing in the supply duct listed in **Problem #93** is 0.075 lbs/ft^3, what is the Velocity Pressure, in inches of water, that will exist in this duct?

Applicable Formula on Page 3-46   Problem Solution on Pages 5-74 & 5-75

Problem #95:

What is the Duct Velocity of the air flowing in a duct under a Velocity Pressure of 2.32 inches of water?

Applicable Formula on Page 3-47   Problem Solution on Page 5-75

Problem #96:

An Industrial Hygienist has measured the Velocity Pressure in an 8-inch duct, and found it to be 0.45 inches of water. What is the Velocity, in fpm, of the gas that is flowing in that duct?

Applicable Formula on Page 3-47   Problem Solution on Page 5-75

Problem #97:

A recently graduated Industrial Hygienist was assigned to measure the Velocity Pressure in a duct carrying air at a Velocity of 25 fps. He reported a value of 0.23 inches of water. Comment on his skill at measuring Velocity Pressures in a duct.

Applicable Formula on Page 3-47   Problem Solution on Page 5-76

Problem #98:

The Total Pressure at a duct exhaust point was measured to be 0.41 inches of water. If the Static Pressure in this duct was 0.01 inches of water, what would you expect the Velocity Pressure to be?

Applicable Formula on Page 3-48   Problem Solution on Page 5-76

Problem #99:

If a Velocity Pressure of 0.55 inches of water, and a Total Pressure of 0.24 inches of water were measured in a duct, at the inlet to the fan for that ventilation system, what would you expect the static pressure to be?

Applicable Formula on Page 3-48   Problem Solution on Page 5-77

Problem #100:

If the fan from **Problem #99** was located 220 feet downstream from its main hood inlet, into which <u>all</u> the air that ultimately passes through the fan is flowing, and if the Frictional Losses in this duct have been determined to be 0.02 inches of water/20 lineal feet of duct, and finally, if the Velocity Pressure (as well as the Duct Velocity) at the fan inlet and at the hood entry are identical, what would you expect the Static Pressure to be at the hood inlet?

Applicable Formula on Page 3-48   Problem Solution on Pages 5-77 & 5-78

Problem #101:

An Air Balancing Technician measured the Static Pressure in a 10-inch diameter duct at two points separated from each other by 75 feet. He found the Static Pressure at the upstream point to be - 0.35 inches of water, and at the downstream point, - 0.41 inches of water. What are the Frictional Losses in this duct, expressed in inches of water/10 lineal feet of duct?

Applicable Formula on Page 3-48   Problem Solution on Page 5-78

Problem #102:

The source of an irritating vapor is located 2.5 feet from the face of a [1 ft]×[3 ft] unflanged, rectangular hood opening. If a capture velocity of 100 fpm is necessary to capture this vapor, what will be the minimum volumetric flow rate, in cfm, required in this system in order to be sure that vapor capture occurs?

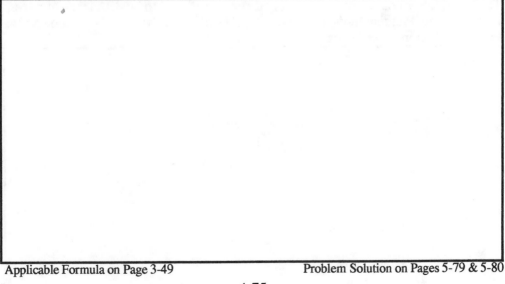

Applicable Formula on Page 3-49   Problem Solution on Pages 5-78 & 5-79

Problem #103:

If the hood opening in **Problem #102** were to be flanged, and if the same hood volumetric flow rate maintained, how far from the hood opening could the irritating vapor source now be located, such that one could still be confident that an adequate vapor capture will continue to occur?

Applicable Formula on Page 3-49   Problem Solution on Pages 5-79 & 5-80

Problem #104:

A source of respirable particles must be mitigated by the use of an exhaust hood. The hood entry slot is to be 3 inches high and 12 inches wide. It is felt that the hood must be flanged, since the source of respirable particles can be no closer than 3.0 feet from the face of the hood slot. The air handling fan in this exhaust system will draw a total of 8,500 cfm. What will the Centerline Velocity of this hood be when it operates under these conditions? If the required Capture Velocity for the respirable particulate material is 2 fps, will this hood system be able to operate satisfactorily and successfully capture the target material?

Applicable Formula on Page 3-49   Problem Solution on Pages 5-80 & 5-81

Problem #105:

The Hood Entry Losses for a simple hood slot were found to be 8.1 inches of water. If the Velocity Pressure in the duct is 15.2 inches of water, what will the Hood Static Pressure be, in inches of water?

Applicable Formula on Page 3-50   Problem Solution on Page 5-81

Problem #106:

What will be the Coefficient of Entry for the hood in **Problem #105**?

Applicable Formula on Page 3-50   Problem Solution on Page 5-81

Problem #107:

The Static Pressure in a hood under investigation was measured to be 12.5 inches of water. The Total Pressure in the duct leading away from this hood, as well as the duct Static Pressure were measured, respectively, to be 6.2 inches of water and - 1.2 inches of water. What are the Hood Entry Losses for this hood?

Applicable Formulae on Pages 3-48 & 3-50   Problem Solution on Pages 5-81 & 5-82

Problem #108:

What will be the Coefficient of Entry for the hood in **Problem #107**?

Applicable Formula on Page 3-50   Problem Solution on Page 5-82

Problem #109:

An Engineer has measured a Static Pressure of - 1.75 inches of water in a 4-inch diameter duct, 12 inches downstream from its unflanged opening. Page 4-12 of **Industrial Ventilation**, identifies the Coefficient of Entry for such a hood opening to be 0.72. Calculate the Volumetric Flow Rate, in cfm, for this system.

Applicable Formula on Page 3-51   Problem Solution on Pages 5-82 & 5-83

Problem #110:

For the hood system described in **Problem #109,** calculate the Duct Velocity, in fpm.

Applicable Formula on Page 3-46   Problem Solution on Pages 5-83 & 5-84

Problem #111:

Finally, for the hood system described in **Problem #109**, what will be the average Duct Velocity Pressure, in inches of water?

Applicable Formula on Page 3-47   Problem Solution on Page 5-84

Problem #112:

What will be the Total Pressure in the Exit Duct of a Hood, that has the following operating characteristics?

Volumetric Flow Rate:	3,000 cfm
Hood Opening:	an unflanged square measuring 6.0 inches on a side
Duct Static Pressure:	- 4.5 inches of water

Applicable Formulae on Pages 3-48 & 3-51   Problem Solution on Pages 5-84 & 5-85

Problem #113:

The Coefficient of Entry for a particular hood was measured to be 0.85. If the Velocity Pressure in the hood is 21.0 inches of water, what will be the Hood Entry Losses, in inches of water?

Applicable Formula on Page 3-52   Problem Solution on Pages 5-85 & 5-86

Problem #114:

Hood Entry Losses for a hood with a Velocity Pressure of 12.3 inches of water were found to be 6.5 inches of water. What is the Coefficient of Entry for this hood?

Applicable Formula on Page 3-52   Problem Solution on Page 5-86

Problem #115:

What is the Hood Entry Loss Factor for the hood in **Problem #114**?

Applicable Formula on Page 3-52   Problem Solution on Pages 5-86 & 5-87

Problem #116:

A compound hood was measured and found to have a Slot Velocity of 3,400 fpm, and a Duct Velocity of 5,600 fpm. What is the Hood Static Pressure, in inches of water, for this compound hood?

Problem #117:

The exhaust duct of a compound hood was measured and found to have a Total Pressure of 1.3 inches of water, and a Static Pressure of - 1.9 inches of water. This Slot Velocity of this hood system was measured and found to be 7,500 fpm. What is the Hood Static Pressure, in inches of water, for this compound hood?

Problem #118:

It was determined that the air-handling fan that serviced a section of a hospital was not moving enough air. To correct this situation, it was determined to change its drive pulley in order to increase its Rotational Speed from 1,450 rpm to 1,750 rpm. If its Discharge Volume is now 10,500 cfm, what was its Discharge Volume before the pulley was changed?

Applicable Formula on Page 3-54   Problem Solution on Page 5-90

Problem #119:

If the Static Discharge Pressure of the fan in **Problem #118** was 4.6 inches of water prior to the change of pulleys, what is its Static Discharge Pressure now?

Applicable Formula on Page 3-54   Problem Solution on Page 5-90

Problem #120:

It was determined that the motor that drives the air-handling fan of **Problem #118** also had to be changed in order to produce the higher fan Rotational Speed. If the previous motor delivered 15 bhp, what must the Brake Horsepower rating of the new drive motor?

Applicable Formula on Page 3-55   Problem Solution on Page 5-91

Problem #121:

A ventilation fan that had been initially selected to deliver 25,000 cfm against 6.25 inches of water Static Pressure, successfully did so at 1,200 rpm. Its Brake Horse-power rating was 50 bhp. Because new facilities are scheduled to be installed in the area serviced by this ventilation system, it has become necessary to increase the Volume Output of this fan by 40%. At what new Rotational Speed, in rpm, must the fan now be operated?

Applicable Formula on Page 3-54   Problem Solution on Page 5-91

Problem #122:

What will be the new Static Discharge Pressure, in inches of water, for the fan in **Problem #121**?

Applicable Formula on Page 3-54   Problem Solution on Page 5-92

Problem #123:

What new Brake Horsepower rating, in bhp, will be required for the motor that drives the fan in **Problem #121**?

Applicable Formula on Page 3-55   Problem Solution on Page 5-92

PROBLEMS

Problem #124:

The Industrial Hygienist of a large manufacturing company has recommended that one of the three exhaust ventilation systems in use at the company's West Coast plant be replaced with a new system, sized so as to be capable of handling the much larger flow volume requirements of the anticipated, expanded manufacturing capability soon to be implemented at this plant. This existing system utilizes a 6.0 inch Diameter fan, and 6.0 inch Diameter ducting throughout. The fan has an established Discharge Capacity of 28,500 cfm. The Industrial Hygienist has calculated the new Discharge Volume requirements of the soon to be modified ventilation system, to be 67,500 cfm. What duct and fan size is this Industrial Hygienist likely to recommend for this new expanded system?

Applicable Formula on Page 3-56   Problem Solution on Page 5-93

Problem #125:

If the fan Static Discharge Pressure (for the existing ventilation system described in **Problem #124**) is 7.0 inches of water, what will be the Static Discharge Pressure of the new system?

Applicable Formula on Page 3-56   Problem Solution on Pages 5-93 & 5-94

4-87

Problem #126:

The Industrial Hygienist identified in **Problem #124** has calculated that the new expanded system will require a fan motor of 150 bhp. What is the Brake Horsepower of the existing system?

Applicable Formula on Page 3-57 Problem Solution on Page 5-94

Problem #127:

A large HVAC system employing a 15.0 inch Diameter fan is characterized by an air Velocity of 150 fps in its 15.0 inch Diameter ducts. If the fan and duct Diameters are each increased by 3.0 inches, what will the new Duct Velocity become?

Work Space Continued on the Next Page

Continuation of Work Space for **Problem #127**

Applicable Formulae on Pages 3-46 & 3-56   Problem Solution on Pages 5-94 thru 5-96

Problem #128:

The air-handler in a ventilation system employs a fan and ducts, each of 8.0 inches Diameter. This system provides a Discharge Air Velocity of 7,525 fpm, and a Discharge Total Pressure of 9.8 inches of water. If the fan in this system is increased in diameter by 50%, what would you forecast the Fan Static Discharge Pressure for the new system to be?

Work Space Continued on the Next Page

Continuation of Work Space for **Problem #128**

Problem #129:

A hood exhaust system uses a 9.0 inch Diameter fan and correspondingly sized ducts. Its fan motor develops 75 bhp. If the fan is decreased in size, a smaller motor could be used. If there is an available 10 bhp motor, by how much must the Diameters of the fan be decreased and have it still be able to satisfy the requirements of this exhaust ventilation system.

Applicable Formula on Page 3-57   Problem Solution on Page 5-97

Problem #130:

What is the Brake Horsepower, in bhp, of a fan that delivers 22,500 cfm at a Mechanical Efficiency of 65%, and at a Fan Total Pressure of 3.4 inches of water?

Applicable Formula on Page 3-58   Problem Solution on Page 5-98

Problem #131:

What is the Mechanical Efficiency of a 20 bhp fan that delivers air at a Velocity of 6,600 fpm into a 12.0 inch Diameter duct at a Fan Total Pressure of 15.0 inches of water?

Applicable Formulae on Pages 3-46 & 3-58   Problem Solution on Pages 5-98 & 5-99

Problem #132:

The inlet duct of a fan had a combined Entry and Friction Loss of 4.5 inches of water, and a Velocity Pressure of 4.2 inches of water. The fan discharge duct has Friction Losses of 4.8 inches of water, and a Velocity Pressure of 5.6 inches of water. Calculate the Fan Total Pressure, in inches of water.

Applicable Formulae on Pages 3-47, 3-48, & 3-58   Problem Solution on Pages 5-99 & 5-100

Problem #133:

The Velocity Pressures at both the fan inlet and fan outlet were measured at 13.4 inches of water. The Static Pressure at the fan inlet was 19.6 inches of water less than the Static Pressure at the fan outlet, and the Fan Total Output Pressure was measured at 25.7 inches of water. What is the Fan Total Pressure?

Work Space Continued on the Next Page

Continuation of Work Space for **Problem #133**

Applicable Formulae on Pages 3-48 & 3-58   Problem Solution on Pages 5-100 & 5-101

Problem #134:

The duct leading into a ventilation fan has a Velocity Pressure of 6.4 inches of water, and a Static Pressure of - 8.3 inches of water. The fan discharge duct has been measured to have a Velocity Pressure of 8.2 inches of water, and a Total Pressure of 16.3 inches of water. Calculate the Fan Static Discharge Pressure.

Applicable Formulae on Pages 3-48 & 3-59   Problem Solution on Page 5-101

Problem #135:

The Fan Static Discharge Pressure of an HVAC fan was determined to be 21.5 inches of water. If the Static Pressures at the fan's inlet and outlet were measured, respectively, to be -12.7 inches of water, and 14.4 inches of water, what was the Velocity Pressure at the fan's inlet?

Applicable Formula on Page 3-59   Problem Solution on Pages 5-101 & 5-102

Problem #136:

Consider the following tabulation of data about the three ducts, AD, BD, & CD, that meet to form duct DD. What must be done to balance this three-into-one duct junction?

Duct Designation	Q, in cfm	V, in fpm	SP, in inches of water
AD	520	3,690	1.26
BD	720	3,590	1.39
CD	950	2,580	1.44

Work Space Continued on the Next Page

Continuation of Work Space for **Problem #136**

Problem #137:

A 1.50 liter glass bottle full of acetone [MW = 58.08, density = 0.791 gms/cc, & Vapor Pressure = 226 mm Hg @ 77°F] falls, during an earthquake, from its position on a shelf, breaking when it hits the floor of a room that has a volume of 1,650 ft^3. If:

   (1)   the conditions in the room are accurately characterized as being at NTP;

   (2)   the room is tightly closed when, and after, this earthquake has occurred; &

   (3)   all the acetone evaporates,

what will the ultimate ambient concentration of acetone be in the room? Is it reasonable to assume that <u>all</u> the acetone will evaporate?

Work Space Continued on the Next Page

Continuation of Work Space for **Problem #137**

Applicable Formulae on Pages 3-17, 3-18, & 3-61   Problem Solution on Pages 5-103 & 5-104

Problem #138:

If the goal is to ventilate the room of **Problem #137** until the acetone concentration is at or below the TLV-TWA concentration, which for acetone is 750 ppm, and if the room's ventilation system has a flow volume capacity of 500 cfm, how long will it take to reduce the acetone concentration to the required level?

Applicable Formula on Page 3-62   Problem Solution on Page 5-105

Problem #139:

After a long day's production run, a Tennessee Moonshiner, who is also a Certified Industrial Hygienist, has filled the bathtub in his 6-foot high, right circular cylindrical redwood tank-storeroom with one batch [one batch ≥ 100 gal] of recently distilled "White Lightning", or pure ethanol. On the following morning, he discovers that exactly 1 gallon of his product has evaporated into the otherwise completely empty and closed tank-storeroom. Ignoring the volume of the bathtub and its remaining contents, what is the diameter, in feet, of this tank-storeroom? You may find it useful to know that the Molecular Weight of ethanol is 46.07 amu; its Density is 0.789 gms/cc; its Vapor Pressure is 44.3 mmHg at NTP, and its formula is CH_3-CH_2-OH. You should also be aware that Moonshiners always keep their storerooms at NTP.

Continuation of Work Space for **Problem #139**

Applicable Formulae on Pages 3-17, & 3-61   Problem Solution on Pages 5-105 thru 5-107

Problem #140:

The CIH of **Problem #139** is very careful to obey all the OSHA requirements that relate to maintaining a safe workplace environment, although he does ignore most other Laws and Statutes. Because of this, he wants to ventilate his vapor filled redwood cylindrical tank-storeroom until the ethanol concentration is at or below its OSHA established PEL-TWA of 1,000 ppm. If his undersized ventilation fan has a capacity of only 25 cfm, for what time period will this entrepreneurial CIH have to ventilate his tank-storeroom to achieve his goal, assuming that he first covers the bathtub so that no more evaporation occurs?

Applicable Formula on Page 3-62   Problem Solution on Page 5-107

Problem #141:

In general, how many "room volumes" of air must be withdrawn from (purged from) a room in order to reduce the concentration of any volatile substance in the ambient air of that room by 90%? by 99%?

Problem #142:

Given the following data set listing workdays without a loss-time accident, tabulated by department for a large factory, determine: (1) the Mean, (2) the Median, (3) the Mode, (4) the Range, (5) the Geometric Mean, (6) the Sample Variance, (7) the Population Variance, (8) the Sample Standard Deviation, (9) the Population Standard Deviation, (10) the Sample Coefficient of Variation, and finally, (11) the Population Coefficient of Variation for this set of data. Every previous analysis of the loss-time accidents in this factory have produced data that has been normally distributed; you may, therefore, assume that this data set too is normally distributed. In the first workspace provided below, it may be helpful to retabulate, in a more useful ordered sequence, the following data set as well as, possibly, some combinations and/or mathematical derivations taken from this data set.

Dept. #	Workdays	Dept. #	Workdays	Dept. #	Workdays
1	85	2	71	3	102
4	43	5	90	6	87
7	55	8	118	9	63
10	62	11	77	12	62
13	95	14	82	15	69

Useful Retabulation of the foregoing Data Set, plus Additional Developed Data:

Retabulated Data on Page 5-110

(1) Mean

Applicable Formula on Page 3-64   Problem Solution on Page 5-111

(2) Median

Applicable Definition on Page 3-64   Problem Solution on Page 5-111

(3) Mode

Applicable Definition on Page 3-65   Problem Solution on Page 5-111

(4) Range

Applicable Definition on Page 3-65   Problem Solution on Page 5-112

(5) **Geometric Mean**

Applicable Formula on Page 3-66   Problem Solution on Page 5-112

(6) **Sample Variance**

Applicable Formula on Page 3-67   Problem Solution on Page 5-112

(7) **Population Variance**

Applicable Formula on Page 3-68   Problem Solution on Page 5-113

(8) Sample Standard Deviation

Applicable Formula on Page 3-69   Problem Solution on Page 5-113

(9) Population Standard Deviation

Applicable Formula on Page 3-70   Problem Solution on Pages 5-113 & 5-114

(10) Sample Coefficient of Variation

Applicable Formula on Page 3-71   Problem Solution on Page 5-114

(11) Population Coefficient of Variation

Applicable Formula on Page 3-71   Problem Solution on Page 5-114

Solutions to Example Problems

PREFACE TO THE PROBLEM SOLUTION SECTION

Although there will be no individualized discussion of the procedure in any of the Problem Solutions that follow, each result — as it is developed and presented in its final "boxed" format — will have been adjusted so as to contain the correct number of significant digits.

In addition, on many of the Problem Solutions in this Section, there will, of necessity, be calculations that develop "intermediate" results — as an example, please refer to **Problem #12**, where there are <u>four</u> separate calculation steps involved in developing the final answer that is asked for in the Problem Statement. In order, for this Problem, these steps are: (1) determining a set of four mass based concentration Standard Values from the given set of four volumetric based Standard Values; (2) determining the overall $TLV_{effective}$ for the mixture; (3) determining what the individual mass based concentrations of each of the four components in this mix would be at this overall $TLV_{effective}$; and (4) finally, determining the four volume based equivalents for each of these mass based concentrations. In every case — and specifically for this Problem — each of the "intermediate" results will also have been reported in an appropriate number of significant digits; <u>however</u>, each subsequent calculation that uses any of these "intermediate" results will employ the "unrounded" number value that has been retained in the math coprocessor of the computer or the calculator that is being used to perform the calculations in this process.

Because of this, any individual who methodically checks <u>each</u> step of any of the following Problem Solutions that have "intermediate" results will almost certainly discover differences in the answers that he or she obtains if, in this overall stepwise process, he or she uses the rounded, rather than the unrounded, "intermediate" values.

As a specific example of this potential difficulty, I would like to discuss two specific calculation steps that are presented in the Solution to **Problem #12**, shown on Pages 5-12 to 5-15 in this Section. The two steps I would like to discuss are both listed on Page 5-13. I would like to start with the determination of C_{mass_4}. This "intermediate" result was calculated to be 173.68 mg/m^3, and was reported — in its rounded form — as 174 mg/m^3. In each subsequent calculation step in the Solution to this Problem, the value of C_{mass_4} <u>appears</u> to have been used in its rounded form; however, such was <u>never</u> the case. Its unrounded equivalent <u>always</u> remained in the math coprocessor (where it had been carried out to a precision of <u>many</u> decimal places) and it was <u>always</u> this value — namely, 173.68+ mg/m^3 — that was used in each subsequent calculation, rather than the indicated rounded 174 mg/m^3 value. The individual who carefully checks out <u>each</u> step in every Problem Solution having "intermediate" results will readily observe the variations that this process has produced.

To continue with this specific example, the expression for the final ratio in the denominator of the overall expression for $TLV_{effective}$, which is taken from the second set of calculations referred to above, is listed as:

$$\frac{0.05}{174}$$

If an individual carries out <u>this</u> mathematical operation, using <u>these</u> values, on <u>any</u> calculator, he or she will obtain as a result, 2.87×10^{-4}. This is obviously different than the listed value of 2.88×10^{-4}; although it is not greatly different. This latter value derives from using the ratio listed below, since it is the <u>unrounded</u> 173.68+ mg/m^3 value that had been maintained in the calculator being used. The ratio actually employed in the math calculation was:

$$\frac{0.05}{173.68 +}$$

It is this second ratio, in contrast to the first one, that produces the slightly different 2.88×10^{-4} value listed above. Analogous slight deviations will likely occur in every Problem Solution where there are any "intermediate" numerical results, and the Reader should be aware of this possibility.

Problem #1:

To solve this problem, we must apply **Equation #1A**, from Page 3-1, thus:

$$TWA = \frac{\sum_{i=1}^{n} T_i C_i}{\sum_{i=1}^{n} T_i} = \frac{T_1 C_1 + T_2 C_2 + \cdots + T_n C_n}{T_1 + T_2 + \cdots + T_n}$$

$$TWA = \frac{[2.0][451] + [3.0][728] + [1.5][619] + [1.5][501]}{2.0 + 3.0 + 1.5 + 1.5}$$

$$TWA = \frac{902.0 + 2,184.0 + 928.5 + 751.5}{8.0} = \frac{4,766.0}{8.0} = 595.75$$

$$\therefore \text{ The TWA} = 596 \text{ ppm}$$

Problem #2:

To solve this problem, we must again apply **Equation #1A**, from Page 3-1.

For this problem, we must assume that the resultant TWA exposure of the Worker was 600 ppm, and then determine the concentration level to which he or she had to have been exposed over the final 15 minutes (= 0.25 hours) of his or her shift, in order to have developed the overall 600 ppm 8-hour TWA. We must consider the final 1.5-hour period as having been made up of <u>two</u> separate and distinct time periods; the first being 1.25 hours (= 75 minutes) — during which the worker was exposed to the 501 ppm level of n-pentane; and the second being 0.25 hours (= 15 minutes) — during which period he or she was exposed to an unknown n-pentane concentration level, to be designated as, "C". It is this concentration level of n-pentane that will have to be determined; thus:

$$\text{TWA} = \frac{\sum_{i=1}^{n} T_i C_i}{\sum_{i=1}^{n} T_i} = \frac{T_1 C_1 + T_2 C_2 + \cdots + T_n C_n}{T_1 + T_2 + \cdots + T_n}$$

$$600 = \frac{[2.0][451] + [3.0][728] + [1.5][619] + [1.25][501] + 0.25C}{2.0 + 3.0 + 1.5 + 1.25 + 0.25}$$

$$600 = \frac{902.0 + 2,184.0 + 928.5 + 626.25 + 0.25C}{8.0} = \frac{4,640.75 + 0.25C}{8.0}$$

$$600 = \frac{4,640.75}{8.0} + \frac{0.25C}{8.0} = 580.09 + \frac{C}{32}$$

$$\frac{C}{32} = 600 - 580.09 = 19.91$$

$$C = [19.91][32] = 637.0$$

∴ The Final 15 minute Concentration Level of n-pentane = 637 ppm

Problem #3:

To solve this problem, we must consider the **Definitions** of the TLV-STEL, from Pages 1-1 & 1-2.

Since the TLV-STEL for n-pentane is 750 ppm, and since this TLV-STEL is a 15-minute Time Weighted Average, and since we are dealing here only with the final 15 minute segment of this Worker's shift, which shift up until these final 15 minutes has had no TLV-STEL violation, it can be concluded that these final 15 minutes must have been the source of the TLV-STEL violation for this work shift; and, therefore, it must have been characterized by a n-pentane exposure level of 750 ppm (or possibly even more than this level).

∴ The Concentration during the final 15 minutes of the shift was 750+ ppm

∴ Obviously, this concentration level of n-pentane would also have produced a TLV-TWA violation — since, from **Problem #2**, we can see that a final 15 minute exposure to only 637 ppm produced a TLV-TWA violation — therefore, an exposure to a higher level [namely, 750+ ppm] would certainly also have produced such a violation.

Problem #4:

To solve this problem, we must again apply **Equation #1A**, from Page 3-1.

For this problem, we shall consider concentrations in parts per billion (ppb), since that is the format in which the concentrations have been provided in the problem (as exposure concentrations for the Technician). Because of this, we must convert the PELs that were provided in the problem statement from the part per million (ppm) units in which they are listed, to part per billion (ppb) units. Remembering that 1,000 ppb = 1 ppm, we can convert the established ozone Permissible Exposure Limits, as follows: the PEL-TWA = 100 ppb, and the PEL-STEL = 300 ppb. We proceed; thus:

$$\text{TWA} = \frac{\sum_{i=1}^{n} T_i C_i}{\sum_{i=1}^{n} T_i} = \frac{T_1 C_1 + T_2 C_2 + \cdots + T_n C_n}{T_1 + T_2 + \cdots + T_n}$$

$$TWA = \frac{[0.5][289] + [1.0][42] + [0.5][93] + [6.0][8]}{0.5 + 1.0 + .05 + 6.0}$$

$$TWA = \frac{144.5 + 42.0 + 46.5 + 48.0}{8.0} = \frac{281.0}{8.0} = 35.13$$

∴ The TWA = 35 ppb

∴ This Technician <u>probably</u> did not experience a PEL-STEL violation; however, his 30 minute diagnostic effort, for which the <u>average</u> exposure was 289 ppb could — at least in principle — have involved one or more 15 minute periods (as subsets of the longer 30 minute period) for which the Time Weighted Average exposure could have exceeded the 300 ppb PEL-STEL.

Problem #5:

To solve this additional twist on **Problem #4**, we must again apply **Equation #1A**, from Page 3-1.

In this case, we must reexamine the final 6 hours of this Technician's day, considering it to be made up of two time periods, as follows: Time Period #1: 2.5 hours @ 222 ppb, and Time Period #2: 3.5 hours @ 8 ppb; thus:

$$TWA = \frac{\sum_{i=1}^{n} T_i C_i}{\sum_{i=1}^{n} T_i} = \frac{T_1 C_1 + T_2 C_2 + \cdots + T_n C_n}{T_1 + T_2 + \cdots + T_n}$$

$$TWA = \frac{[0.5][289] + [1.0][42] + [0.5][93] + [3.5][8] + [2.5][222]}{0.5 + 1.0 + .05 + 3.5 + 2.5}$$

$$TWA = \frac{144.5 + 42.0 + 46.5 + 28.0 + 555.0}{8.0} = \frac{816.0}{8.0} = 102.0$$

∴ The TWA = 102 ppb > 0.1 ppm

∴ Clearly, this Technician had experienced an 8-hour TWA ozone exposure that exceeded the established OSHA PEL-TWA of 0.1 ppm — 102 ppb > 0.1 ppm

Problem #6:

To solve this Problem, we must again apply **Equation #1A**, from Page 3-1.

The analysis of this problem is comparable to those for the previous two problems; however, since one of the timed average ethylene oxide exposures listed in this problem statement has been provided in a "less than" format, we must think carefully about the implications of this. Specifically, what this means is that the ethylene oxide concentration, during that 2.9 hour time period, would be more accurately and usefully presented in the following format:

$$0.0 \text{ ppm} \le \text{ethylene oxide concentration} < 0.4 \text{ ppm}$$

As a consequence of this, the final result will, of necessity, also have to be in a slightly different format from any of the previous results. To develop it, we will have to consider two different situations: (1) a maximum possible TWA value, and (2) a minimum possible TWA value. The maximum TWA value will be developed by assuming that this time segment actually involved a 2.9 hour exposure <u>at a</u> <u>concentration</u> <u>of 0.4 ppm</u>. Clearly, this is a situation that is <u>worse</u> than what might otherwise have been the worst possible case. The number that will result from this calculation will identify a TWA exposure that will be <u>higher</u> than what was the true value for the case. The minimum TWA value, on the other hand, will be developed by assuming that this same 2.9 hour exposure time segment was at a 0.0 ppm ethylene oxide concentration level. This 0.0 ppm concentration level clearly is less than 0.4 ppm; it is also certainly equal to, or less than, the actual "indeterminable" concentration value ("indeterminable", because the Minimum Detection Limit for the analyzer that was used was 0.4 ppm) for this segment; thus:

$$TWA = \frac{\sum\limits_{i=1}^{n} T_i C_i}{\sum\limits_{i=1}^{n} T_i} = \frac{T_1 C_1 + T_2 C_2 + \cdots + T_n C_n}{T_1 + T_2 + \cdots + T_n}$$

$$TWA_{max} = \frac{[2.2][0.7] + [1.4][1.2] + [1.5][1.6] + [2.9][0.4]}{2.2 + 1.4 + 1.5 + 2.9}$$

$$TWA_{max} = \frac{1.54 + 1.68 + 2.40 + 1.16}{8.0} = \frac{6.78}{8.0} = 0.85$$

$$TWA_{min} = \frac{[2.2][0.7] + [1.4][1.2] + [1.5][1.6] + [2.9][0]}{2.2 + 1.4 + 1.5 + 2.9}$$

$$\text{TWA}_{min} = \frac{1.54 + 1.68 + 2.40 + 0.00}{8.0} = \frac{5.62}{8.0} = 0.70$$

> ∴ The Requested Result is 0.7 ppm ≤ TWA< 0.9 ppm

∴ Clearly, the exposure is less than the 8-hour PEL-TWA of 1.0 ppm; however, it is more than the 0.5 ppm Action Level. As such, it indicates the existence of a potential bad situation that a prudent Professional should examine, from a preventative perspective, in order to prevent any future PEL violation. It would have to be reported as an exposure that exceeded the Action Level for the Worker involved.

Problem #7:

To solve this problem, we must again apply **Equation #1A**, from Page 3-1.

There are aspects to this problem that are identical to those in **Problem #6**, namely, in the area of having certain timed average exposures to styrene, reported in a "less than" format. The approach to this problem, therefore, will be analogous to the one employed for **Problem #6**. In this case, however, since we have been given the overall 8-hour TWA, we shall have to evaluate the situation by considering the two different time intervals — namely, the 7.33 hour segment in the Production Area (identified subsequently as Time Interval #1), and the 0.67 hour segment in the "Low Level Room" (identified subsequently as Time Interval #2) — and perform the analysis from two different perspectives. These two perspectives are as follows:

1. In order to have developed the overall 48 ppm TWA styrene exposure, the highest concentration level of styrene vapors that could have existed in the Production Area must have corresponded to the lowest styrene concentration level that could have existed in the "Low Level Room". These two first scenario concentrations will be designated as C_{max} in the Production Area, and 0.0 ppm in the "Low Level Room".

2. Conversely, the lowest concentration level of styrene vapors that could have existed in the Production Area must correspond to the highest styrene concentration level that could have existed in the "Low Level Room". These two second scenario concentrations will be designated as C_{min} in the Production Area, and 0.1 ppm in the "Low Level Room".

Thus we view this situation, as follows:

Time Interval #	Duration	Location	First Scenario Concentrations	Second Scenario Concentrations
1	7.33 hours	Production Area	C_{max}	C_{min}
2	0.67 hours	"Low Level Room"	0.0 ppm	0.1 ppm
Total	8.00 hours			

With this approach, we can proceed; thus:

$$TWA = \frac{\sum_{i=1}^{n} T_i C_i}{\sum_{i=1}^{n} T_i} = \frac{T_1 C_1 + T_2 C_2 + \cdots + T_n C_n}{T_1 + T_2 + \cdots + T_n}$$

$$48 = \frac{[0.67][0.1] + 7.33 C_{min}}{0.67 + 7.33} = \frac{0.067 + 7.33 C_{min}}{8.00}$$

$$48 = \frac{0.067}{8.00} + \frac{7.33 C_{min}}{8.00} = 0.008 + \frac{22 C_{min}}{[8.00][3]} = 0.008 + 0.917 C_{min}$$

$$0.917 C_{min} = 48 - 0.008 = 47.992$$

$$C_{min} = \frac{47.992}{0.917} = 52.355$$

$$48 = \frac{[0.67][0.0] + 7.33 C_{max}}{0.67 + 7.33} = \frac{0.00 + 7.33 C_{max}}{8.00}$$

$$48 = \frac{0.00}{8.00} + \frac{7.33 C_{max}}{8.00} = 0.00 + \frac{22 C_{max}}{[8.00][3]} = 0.917 C_{max}$$

$$0.917 C_{max} = 48$$

$$C_{max} = \frac{48}{0.917} = 52.364$$

As you can see, these two calculations have been carried out to the nearest 0.001 ppm. At this level, there is a mathematical difference between the two styrene exposure levels (ie. the $C_{min} = 52.355$ ppm, and the $C_{max} = 52.364$ ppm). This difference is, however, <u>completely inconsequential</u>; thus, it will be ignored.

∴ Production Area Styrene Ambient Concentration = 52 ppm.

∴ Again, in this case, the competent Professional would probably focus some of his or her attention on this matter, since this scenario is close to being a PEL violation, with dangerously high styrene levels in the Production Area.

Problem #8:

To solve this problem, we must again apply **Equation #1A**, from Page 3-1.

Clearly, the average worker's 8-hour TWA exposure to CO_2 is at the 8,400 ppm (84% of the PEL-TWA, which is 10,000 ppm). To determine the Cold Room concentration of CO_2, we must proceed; thus:

$$TWA = \frac{\sum_{i=1}^{n} T_i C_i}{\sum_{i=1}^{n} T_i} = \frac{T_1 C_1 + T_2 C_2 + \cdots + T_n C_n}{T_1 + T_2 + \cdots + T_n}$$

$$8,400 = \frac{6.0 C_{Cold\ Room} + [2.0][425]}{6.0 + 2.0} = \frac{6.0 C_{Cold\ Room} + 850}{8.0}$$

$$8,400 = \frac{6.0 C_{Cold\ Room}}{8.0} + \frac{850}{8.0} = 0.75 C_{Cold\ Room} + 106.25$$

$$0.75 C_{Cold\ Room} = 8,400 - 106.25 = 8,293.75$$

$$C_{Cold\ Room} = \frac{8,293.75}{0.75} = 11,058.3$$

∴ Cold Room Average CO_2 Concentration = 11,058 ppm.

Problem #9:

Again, to solve this problem, we must apply **Equation #1A**, from Page 3-1.

This problem is both straightforward and analogous to all of the preceding problems, except that in this case, we have been asked for, and will focus on, two unknown time intervals, the first of which shall be referred to as t_{max}, which will represent the maximum number of hours in each workday that the Cold Storage Plant workers will be permitted to work in the Cold Rooms (which will have a minimum background CO_2 level of 15,000 ppm) — under the new CO_2 exposure requirements, as they have been defined by the Safety Manager — while the second, comprising the balance of the 8-hour workday, will be identified as $(8 - t_{max})$; thus:

$$TWA = \frac{\sum_{i=1}^{n} T_i C_i}{\sum_{i=1}^{n} T_i} = \frac{T_1 C_1 + T_2 C_2 + \cdots + T_n C_n}{T_1 + T_2 + \cdots + T_n}$$

$$5,000 = \frac{15,000 t_{max} + 425(8 - t_{max})}{t_{max} + (8 - t_{max})} = \frac{15,000 t_{max} - 425 t_{max} + [425][8]}{t_{max} + 8 - t_{max}}$$

$$5,000 = \frac{14,575 t_{max} + 3,400}{8} = \frac{14,575 t_{max}}{8} + \frac{3,400}{8} = 1,821.9 t_{max} + 425$$

$$1,821.9 t_{max} = 5,000 - 425 = 4,575$$

$$t_{max} = \frac{4,575}{1,821.9} = 2.51$$

∴ The Maximum Time Period out of each Workday that a Worker can be Permitted to be in the Cold Room = 2.5 hours = 2 hours and 30 minutes.

Problem #10:

To solve this problem, we must apply **Equation #1B**, from Page 3-2.

This problem is very straightforward and can be solved; thus:

$$\%TLV = 100 \left[\sum_{i=1}^{n} \frac{TWA_i}{TLV_i} \right] = 100 \left[\frac{TWA_1}{TLV_1} + \frac{TWA_2}{TLV_2} + \cdots + \frac{TWA_n}{TLV_n} \right]$$

$$\%TLV = 100 \left[\frac{12}{50} + \frac{17}{100} + \frac{55}{200} + \frac{91}{400} \right] = 100[0.24 + 0.17 + 0.28 + 0.23]$$

$$\%TLV = 100[0.91] = 91.25$$

These Workers experience just over 91% of the established TLV for these four volatile materials.

$$\therefore \ \%TLV = 91\%$$

Problem #11:

To solve this problem, we must again apply **Equation #1B**, from Page 3-2.

For this problem, we will perform two different sets of calculations, only one of which has any validity. In order, these will be a %TLV-TWA calculation, which is the valid determination, and a %TLV-STEL calculation, which is invalid and only useful as an <u>approximation</u> of potential problems (the stated dosages are 8-hour TWAs, whereas the %TLV-STEL Standard is a 15-minute TWA, thus the two are completely mutually inconsistent). We must also again deal with differences in the listed concentration units. The average exposures are given in parts per billion, while the ACGIH Standards are all listed in parts per million. It will be easiest to convert everything into parts per billion; therefore, we have the Employee's average 8-hour TWA chlorine exposure = 420 ppb, and his 8-hour TWA chlorine dioxide counterpart = 80 ppb. Let us start with the %TLV-TWA determination, using **Equation #1B** from Page 3-2; thus:

$$\%TLV - TWA = 100\left[\sum_{i=1}^{n}\frac{TWA_i}{TLV - TWA_i}\right] = 100\left[\frac{TWA_1}{TLV - TWA_1} + \frac{TWA_2}{TLV - TWA_2}\right]$$

$$\%TLV - TWA = 100\left[\frac{420}{500} + \frac{80}{100}\right] = 100[0.84 + 0.80]$$

$$\%TLV - TWA = 100[1.64] = 164$$

Now let us develop the "invalid" %TLV-STEL, that we may possibly be able to use as a measure of how good or bad things might be from the perspective of the Short Term Exposure Limit, again using **Equation #1B** from Page 3-2; thus:

$$\%TLV - STEL = 100\left[\sum_{i=1}^{n}\frac{TWA_i}{STEL_i}\right] = 100\left[\frac{TWA_1}{STEL_1} + \frac{TWA_2}{STEL_2}\right]$$

$$\%TLV - STEL = 100\left[\frac{420}{1000} + \frac{80}{300}\right] = 100[0.4200 + 0.2667]$$

$$\%\mathrm{TLV-STEL} = 100[0.6867] = 68.67$$

To put both of the foregoing results into a proper perspective, the following can be said — clearly, there is a problem with this Kraft Pulp Bleaching Mill's ambient environment, particularly on heavy bleaching days. The Employee identified in this problem received an exceptionally dangerous exposure to these two very irritating chemicals. The resultant fictitious %TLV-STEL of 69% appears to be all right; however, the different time bases — as stated above — make this value useful only as a general indication.

∴ This Worker's %TLV (based on the 8-hour TLV-TWA Standard) = 164%

∴ Clearly this Worker's employer, the Kraft Pulp Mill Company, is in violation — severely so — of the established TLV-TWA Standard, as it applies to an environment that contains, simultaneously, both of these two very hazardous, toxic, and irritating chemicals. It is also interesting to note that <u>neither</u> of these two chemical's individual PEL-TWAs has been exceeded (at least as defined in this problem), when each is considered by itself.

Problem #12:

To solve this problem, we must apply a total of three of the listed relationships, and do so in the sequence listed, as follows: **Equation #2C**, from Page 3-5, then **Equation #1C** form Page 3-3, and finally **Equation #2A** from Page 3-4.

Again, note that this relationship is an approximation, one that relies upon the assumption that the overall composition of the vapor space above a liquid mixture of volatile solvents has the same composition as does the solution, itself. Such a situation would virtually <u>never</u> exist in the real world; however, it is the basis for the approximation developed in this relationship. The first thing that must be done in obtaining a solution to this problem is to convert the four volume based TLV-TWAs that have been provided in the problem statement, into their mass based equivalents, using **Equation #2C** from Page 3-5; thus:

$$C_{mass_i} = \frac{MW_i}{24.45}\left[C_{vol_i}\right] \quad @ \text{ NTP Conditions}$$

1. For Freon 11:

$$C_{mass_1} = \frac{137.38}{24.45}[1,000] = \frac{137,380}{24.45} = 5,618.81$$

$$\mathrm{TLV\text{-}TWA}_{\text{Freon 11}} = 5,619 \text{ mg/m}^3$$

2. For Freon 113: $\quad C_{mass_2} = \dfrac{187.38}{24.45}[1,000] = \dfrac{187,380}{24.45} = 7,663.80$

$\text{TLV-TWA}_{\text{Freon 113}} = 7,664 \text{ mg/m}^3$

3. For methyl chloroform: $\quad C_{mass_3} = \dfrac{133.41}{24.45}[350] = \dfrac{46,693.50}{24.45} = 1,909.75$

$\text{TLV-TWA}_{\text{methyl chloroform}} = 1,910 \text{ mg/m}^3$

4. For methylene chloride: $\quad C_{mass_4} = \dfrac{84.93}{24.45}[50] = \dfrac{4,246.50}{24.45} = 173.68$

$\text{TLV-TWA}_{\text{methlyene chloride}} = 174 \text{ mg/m}^3$

Next, we must next apply **Equation #1C**, from Page 3-3, since we have now converted all the TLV concentrations from their volumetric basis (ie. ppm) to their corresponding mass basis (ie. mg/m^3), as required for incorporation into this Equation. We can now determine the overall ambient air concentration (in the mass based units) of this vaporous mix of all four degreasing solvent concentrations (in milligrams of solvent vapor per cubic meter of air) as they would hypothetically exist in equilibrium with the air. Remember that this result is an approximation that relies upon the unlikely situation that the vapor and the liquid compositions are identical; thus:

$$TLV_{\text{effective}} = \left| \dfrac{1}{\displaystyle\sum_{i=1}^{n} \dfrac{f_i}{TLV_i}} \right| = \left[\dfrac{1}{\dfrac{f_1}{TLV_1} + \dfrac{f_2}{TLV_2} + \cdots + \dfrac{f_n}{TLV_n}} \right]$$

$$TLV_{\text{effective}} = \left| \dfrac{1}{\dfrac{0.25}{5,619} + \dfrac{0.55}{7,664} + \dfrac{0.15}{1,910} + \dfrac{0.05}{174}} \right|$$

$$TLV_{\text{effective}} = \left| \dfrac{1}{4.45 \times 10^{-5} + 7.18 \times 10^{-5} + 7.85 \times 10^{-5} + 2.88 \times 10^{-4}} \right|$$

$$TLV_{effective} = \frac{1}{4.83 \times 10^{-3}} = 2,071.73$$

∴ The Effective Overall TLV for this Mixture of Solvent Vapors = 2,072 mg/m³.

Now, we can determine — for this overall 4-solvent, mass based concentration value (expressed in mg/m³) — the concentration of each of the individual solvent component (also expressed in mg/m³), simply by applying the solution weight ratios, which were given in the initial problem statement, to this just determined total vapor concentration.

1. For Freon 11: $C_{mass_1} = [2,071.73][0.25] = 517.93$

$C_{Freon\ 11} = 518$ mg/m³

2. For Freon 113: $C_{mass_2} = [2,071.73][0.55] = 1,139.45$

$C_{Freon\ 113} = 1,139$ mg/m³

3. methyl chloroform: $C_{mass_3} = [2,071.73][0.15] = 310.76$

$C_{methyl\ chloroform} = 311$ mg/m³

4. methylene chloride: $C_{mass_4} = [2,071.73][0.05] = 103.59$

$C_{methylene\ chloride} = 104$ mg/m³

With these four individual mass based concentration values, we can finally apply the last necessary relationship, namely, **Equation #2A** from Page 3-4, thereby converting these mass based values (expressed in mg/m³) to their equivalent volume based concentrations (expressed in ppm); thus:

$$C_{vol_i} = \frac{24.45}{MW_i}\left[C_{mass_i}\right] \quad @\ NTP\ Conditions$$

1. For Freon 11: $C_{vol_1} = \frac{24.45}{137.38}[517.93] = \frac{12,663.45}{137.38} = 92.18$

$C_{Freon\ 11} = 92$ ppm

2. For Freon 113: $C_{vol_2} = \dfrac{24.45}{187.38}[1,139.45] = \dfrac{27,859.59}{187.38} = 148.68$

$C_{Freon\ 113} = 149$ ppm

3. For methyl chloroform: $C_{vol_3} = \dfrac{24.45}{133.41}[310.76] = \dfrac{7,598.07}{133.41} = 56.95$

$C_{methyl\ chloroform} = 57$ ppm

4. For methylene chloride: $C_{vol_4} = \dfrac{24.45}{84.93}[103.59] = \dfrac{2,532.69}{84.93} = 29.82$

$C_{methylene\ chloride} = 30$ ppm

Finally, to summarize this set of data, which is the set that was asked for in the problem statement, we can produce the following tabulation of the ambient concentrations of the four degreasing solvents.

∴	Freon 11 Concentration	=	518 mg/m^3
		=	92 ppm
	Freon 113 Concentration	=	1,139 mg/m^3
		=	149 ppm
	Methyl Chloroform Concentration	=	311 mg/m^3
		=	57 ppm
	Methylene Chloride Concentration	=	104 mg/m^3
		=	30 ppm

Problem #13:

This problem is very similar to **Problem #12**, except that this one involves only two, rather than four, volatile chemicals. Like its predecessor, it will require the use of the same three listed relationships, and in the same sequence, namely: **Equation #2C** on Page 3-5, **Equation #1C** on Page 3-3, and finally **Equation #2A** on Page 3-4.

Again, we must first make the conversion of the two volume based TLV-TWAs (expressed in ppm) to their mass based equivalents (expressed in mg/m^3), using **Equation #2C** from Page 3-5; thus:

$$C_{mass_i} = \frac{MW_i}{24.45}\left[C_{vol_i}\right] \quad @ \text{ NTP Conditions}$$

1. For cellosolve:
$$C_{mass_1} = \frac{90.12}{24.45}[200] = \frac{18,024}{24.45} = 737.18$$

$$\text{TLV-TWA}_{cellosolve} = 737 \text{ mg/m}^3$$

2. For t-butyl alcohol:
$$C_{mass_2} = \frac{74.12}{24.45}[100] = \frac{7412}{24.45} = 303.15$$

$$\text{TLV-TWA}_{t\text{-butyl alcohol}} = 303 \text{ mg/m}^3$$

Next, we must use these mass based values to determine the Effective TLV for the mixture of solvent vapors that might exist above the solution, again — as with the previous problem — using **Equation #1C** from Page 3-3; thus:

$$TLV_{effective} = \left|\frac{1}{\displaystyle\sum_{i=1}^{n}\frac{f_i}{TLV_i}}\right| = \left[\frac{1}{\dfrac{f_1}{TLV_1} + \dfrac{f_2}{TLV_2}}\right]$$

$$TLV_{effective} = \left|\frac{1}{\dfrac{0.65}{737.18} + \dfrac{0.35}{303.15}}\right| = \left[\frac{1}{8.82 \times 10^{-4} + 1.15 \times 10^{-3}}\right]$$

$$TLV_{effective} = \frac{1}{2.04 \times 10^{-3}} = 491.09$$

∴ the effective overall TLV for this mixture of solvent vapors = 491 mg/m³.

It will now be quite easy to develop the mass based concentrations for the two solvents in this solution, simply by multiplying the overall concentration value (the one that was just determined above) by the solution weight fractions that were provided in the problem statement; thus:

1. For cellosolve:
$$C_{mass_1} = [491.09][0.65] = 319.21$$

$$C_{cellosolve} = 319 \text{ mg/m}^3$$

2. For t-butyl alcohol: $\quad C_{mass_2} = [491.09][0.35] = 171.88$

$$C_{\text{t-butyl alcohol}} = 172 \text{ mg/m}^3$$

With these two mass based concentrations, we can now apply **Equation #2A** from Page 3-4, to obtain the final pieces of information requested in the problem statement; thus:

$$C_{vol_i} = \frac{24.45}{MW_i}\left[C_{mass\,i}\right] \quad @ \text{ NTP Conditions}$$

1. For cellosolve: $\quad C_{vol_1} = \frac{24.45}{90.12}[319.21] = \frac{7,804.64}{90.12} = 86.60$

$$C_{\text{cellosolve}} = 87 \text{ ppm}$$

2. For t-butyl alcohol: $\quad C_{vol_2} = \frac{24.45}{74.12}[171.88] = \frac{4,202.50}{74.12} = 56.70$

$$C_{\text{t-butyl alcohol}} = 57 \text{ ppm}$$

The final set of data that was requested in the statement of the problem is a summary of the concentrations of these two volatile solvents, as they might exist above the solution.

∴	cellosolve Concentration	=	319 mg/m^3
		=	87 ppm
	t-butyl alcohol Concentration	=	172 mg/m^3
		=	57 ppm

Problem #14:

This problem is also very similar to **Problem #'s 12 & 13**, except that this one involves three, rather than four or two, different volatile chemicals. Like its predecessors, it will require the use of the same three relationships, and in the same sequence, namely: **Equation #2C** from Page 3-5, **Equation #1C** from Page 3-3, and finally **Equation #2A** from Page 3-4.

We must first make the conversion of the three volume based TLV-TWAs (expressed in ppm) for each of these refrigerants, to their mass based equivalents (expressed in mg/m^3), using **Equation #2C** from Page 3-5; thus:

$$C_{mass_i} = \frac{MW_i}{24.45}\left[C_{vol\,i}\right] \quad @ \text{ NTP Conditions}$$

1. For Freon 12:
$$C_{mass_1} = \frac{120.92}{24.45}[1,000] = \frac{120,920}{24.45} = 4,945.60$$

$$\text{TLV-TWA}_{Freon\,12} = 4,946 \text{ mg/m}^3$$

2. For Freon 21:
$$C_{mass_2} = \frac{102.92}{24.45}[10] = \frac{1,029.20}{24.45} = 42.09$$

$$\text{TLV-TWA}_{Freon\,21} = 42 \text{ mg/m}^3$$

3. For Freon 112:
$$C_{mass_3} = \frac{170.92}{24.45}[500] = \frac{85,460.00}{24.45} = 3,495.30$$

$$\text{TLV-TWA}_{Freon\,112} = 3,495 \text{ mg/m}^3$$

Next, these three mass based TLV-TWA values can be readily used to determine the Effective TLV for the mixture, using **Equation #1C** from Page 3-3; thus:

$$TLV_{effective} = \left|\frac{1}{\sum\limits_{i=1}^{n}\dfrac{f_i}{TLV_i}}\right| = \left[\frac{1}{\dfrac{f_1}{TLV_1} + \dfrac{f_2}{TLV_2} + \dfrac{f_n}{TLV_3}}\right]$$

$$TLV_{effective} = \left|\frac{1}{\dfrac{0.40}{4,945.60} + \dfrac{0.20}{42.09} + \dfrac{0.40}{3,495.30}}\right|$$

$$TLV_{effective} = \left|\frac{1}{8.09 \times 10^{-5} + 4.75 \times 10^{-3} + 1.14 \times 10^{-4}}\right|$$

$$TLV_{effective} = \frac{1}{4.95 \times 10^{-3}} = 202.16$$

This is the initial piece of information that was asked for in the problem statement.

∴ The Effective Overall TLV for this Mixture of Three Different Refrigerant Vapors = 202 mg/m³.

We must now calculate the three mass based concentrations that would make up an overall TLV level concentration of these three refrigerants. As with the previous two problems, this is easily accomplished simply by multiplying the overall mixture TLV obtained above by the weight percentages of each of the three refrigerants, as they were provided in the problem statement; thus:

1. For Freon 12: $\quad C_{mass_1} = [202.16][0.40] = 80.86$

$C_{Freon\ 12} = 81$ mg/m³

2. For Freon 21: $\quad C_{mass_2} = [202.16][0.20] = 40.43$

$C_{Freon\ 21} = 40.4$ mg/m³

3. For Freon 112: $\quad C_{mass_3} = [202.16][0.40] = 80.86$

$C_{Freon\ 112} = 81$ mg/m³

With these three mass based concentrations, we can now apply **Equation #2A** from Page 3-4, to obtain the final pieces of information requested in the problem statement; thus:

$$C_{vol_i} = \frac{24.45}{MW_i}\left[C_{mass\ i}\right] \quad @ \ NTP \ Conditions$$

1. For Freon 12: $\quad C_{vol_1} = \frac{24.45}{120.92}[80.86] = \frac{1,977.12}{120.92} = 16.35$

$C_{Freon\ 12} = 16$ ppm

2. For Freon 21:

$$C_{vol_2} = \frac{24.45}{102.92}[40.43] = \frac{988.56}{102.92} = 9.61$$

$$C_{Freon\ 21} = 9.6\ ppm$$

3. For Freon 112:

$$C_{vol_3} = \frac{24.45}{170.92}[80.86] = \frac{1,977.12}{170.92} = 11.57$$

$$C_{Freon\ 112} = 12\ ppm$$

We now can tabulate the final set of data asked for in the problem statement.

$\therefore$	Freon 12 concentration	=	81 mg/m3
		=	16 ppm
	Freon 21 concentration	=	40.4 mg/m3
		=	9.6 ppm
	Freon 112 concentration	=	81 mg/m3
		=	12 ppm

Problem #15:

The solution to this problem relies simply on the results obtained in **Problem #14**.

Since the concentration of Freon 21 (ie. the material to which the U. S. Navy's highly specific, bulkhead mounted ambient air analyzer responds), at a time when the overall Effective TLV for this mixture has been achieved, is 9.6 ppm [see the results of **Problem #14**], this is the concentration upon which the Alarm Settings for this analyzer will be based. Remembering, too, that the precision of this analyzer is $\pm\ 0.05$ ppm, we can proceed, thus:

1. For the Alert Alarm:

$$C_{Alert\ Alarm} = (0.60)(9.61) = 5.76$$

Alert Alarm Setting = 5.7 ppm

2. For the Evacuate Alarm:

$$C_{Evacuate\ Alarm} = (0.90)(9.61) = 8.64$$

Evacuate Alarm Setting = 8.6 ppm

With these settings, either alarm will be energized — considering the precision of the analyzer in question — at an ambient concentration that is <u>either</u> *just slightly below* the target concentration for that alarm [ie. the Alert Alarm], <u>or</u> *mostly just slightly below* BUT *possibly barely above* the target concentration for that alarm [ie. the Evacuate Alarm, where a concentration 10 ppb higher than the desired concentration — 8.65 ppm vs. 8.64 ppm — could be achieved before the alarm sounded]. For confirmation of this situation, please see the tabulation that follows:

Alarm Type	Alarm Setting	Concentration Range at which an Alarm will Sound	Target Concentration
Alert	5.7 ppm	5.65 ppm $\leq$ Conc. $_{\text{Freon 21}}$ $\leq$ 5.75 ppm	5.76 ppm
Evacuate	8.6 ppm	8.55 ppm $\leq$ Conc. $_{\text{Freon 21}}$ $\leq$ 8.65 ppm	8.64 ppm

$\therefore$	Alert Alarm Setting	=	5.7 ppm
	Evacuate Alarm Setting	=	8.6 ppm

Problem #16:

To solve this problem, we must apply **Equation #2A** from Page 3-4, thus:

$$C_{vol} = \frac{24.45}{MW}[C_{mass}] \quad @ \text{ NTP Conditions}$$

For the MAK-TWA:
$$C_{vol} = \frac{24.45}{100.12}[0.8] = \frac{19.56}{100.12} = 0.20$$

$$\text{MAK-TWA}_{glutaraldehyde} = 0.20 \text{ ppm}$$

$\therefore$ MAK-TWA $_{glutaraldehyde}$ = 0.2 ppm

Problem #17:

To solve this problem, we must apply **Equation #2B** from Page 3-4, thus:

$$C_{vol} = \frac{22.41}{MW}[C_{mass}] \quad @ \text{ STP Conditions}$$

1. For the PEL-STEL:
$$C_{vol} = \frac{22.41}{72.11}[735] = \frac{16,471.35}{72.11} = 228.42$$

$$\text{PEL-STEL}_{tetrahydrofuran} = 228 \text{ ppm}$$

2. For the PEL-TWA:

$$C_{vol} = \frac{22.41}{72.11}[590] = \frac{13,221.90}{72.11} = 183.36$$

PEL-TWA $_{tetrahydrofuran}$ = 183 ppm

$$\therefore \text{PEL-STEL}_{tetrahydrofuran} = 228 \text{ ppm}$$
$$\text{PEL-TWA}_{tetrahydrofuran} = 183 \text{ ppm}$$

Problem #18:

To solve this problem, we must apply **Equation #2C** from Page 3-5, thus:

$$C_{mass} = \frac{MW}{24.45}[C_{vol}] \quad @ \text{ NTP Conditions}$$

1. For the PEL-TWA:

$$C_{mass} = \frac{44.05}{24.45}[1.0] = \frac{44.05}{24.45} = 1.80$$

PEL-TWA $_{ethylene\ oxide}$ = 1.8 mg/m^3

2. For the Action Level:

$$C_{mass} = \frac{44.05}{24.45}[0.5] = \frac{22.03}{24.45} = 0.90$$

Action Level $_{ethylene\ oxide}$ = 0.9 mg/m^3

$$\therefore \text{PEL-TWA}_{ethylene\ oxide} = 1.8 \text{ mg/m}^3$$
$$\text{Action Level}_{ethylene\ oxide} = 0.9 \text{ mg/m}^3$$

Problem #19:

To solve this problem, we must apply **Equation #2D** from Page 3-5, thus:

$$C_{mass} = \frac{MW}{22.41}[C_{vol}] \quad @ \text{ STP Conditions}$$

1. For the PEL-TWA:
$$C_{mass} = \frac{78.11}{22.41}[1.0] = \frac{78.11}{22.41} = 3.49$$

PEL-TWA $_{benzene}$ = 3.5 mg/m^3

2. For the PEL-STEL:
$$C_{mass} = \frac{78.11}{22.41}[5.0] = \frac{390.55}{22.41} = 17.43$$

PEL-STEL $_{benzene}$ = 17.4 mg/m^3

3. For the PEL-C:
$$C_{mass} = \frac{78.11}{22.41}[25.0] = \frac{1,952.75}{22.41} = 87.14$$

PEL-C $_{benzene}$ = 87.1 mg/m^3

∴ PEL-TWA $_{benzene}$ = 3.5 mg/m^3
PEL-STEL $_{benzene}$ = 17.4 mg/m^3
PEL-C $_{benzene}$ = 87.1 mg/m^3

Problem # 20:

The solution to this problem requires the application of **Equation #3A**, from Page 3-6, thus:

$$TLV_{quartz} = \frac{10\,mg\,/\,m^3}{\%\,RQ + 2}$$

For the % Respirable Quartz: $0.22 = \dfrac{10}{\%\,RQ + 2}$ & then transposing:

$$\%RQ + 2 = \frac{10}{0.22} = 45.45$$

$$\%RQ = 45.45 - 2 = 43.45$$

∴ The Percent Respirable Quartz Level in this River Gravel = 43%

Problem #21:

This problem clearly requires the use of **Equation #3C**, from Page 3-7, thus:

$$TLV_{mix} = \frac{10mg/m^3}{\%Q + [2][\%C] + [2][\%T] + 2}$$

In order to apply this formula, we must first determine the percentages of each of the three types of silica that make up the material mix being handled. We have been provided with a characteristic "mix ratio" for this material — one that identifies the proportions of each of the three basic silica components in the mix being evaluated, thus; we can see:

$$18 + 13 + 9 = 40$$

We have been further advised of the fact that this mix contains 80% silica (ie. this 80% consists of the three basic types of silica — quartz, cristobalite, & tridymite); thus, for the total material mix being considered, we can say:

1. For the percent quartz: $\qquad\qquad \%Q = \frac{18}{40}[80] = 36\%$

2. For the percent cristobalite: $\qquad\qquad \%C = \frac{13}{40}[80] = 26\%$

3. For the percent tridymite: $\qquad\qquad \%T = \frac{9}{40}[80] = 18\%$

and finally applying **Equation #3C**, as listed at the top of this page:

$$TLV_{mix} = \frac{10mg/m^3}{\%Q + [2][\%C] + [2][\%T] + 2}$$

$$TLV_{mix} = \frac{10}{36 + [2][26] + [2][18] + 2}$$

$$TLV_{mix} = \frac{10}{36 + 52 + 36 + 2} = \frac{10}{126} = 0.079$$

∴ The TLV $_{mix}$ = 0.08 mg/m³.

Problem #22:

The solution to this problem requires the use of Boyle's Law, **Equation #4A**, from Page 3-8, thus:

$$P_1 V_1 = P_2 V_2$$

$$[175]V_1 = [1,013.25][100] = 1.01325 \times 10^5$$

$$V_1 = \frac{1.01324 \times 10^5}{175} = 579$$

$\therefore V_1 = 579$ liters

Problem #23:

The solution to this problem also requires the use of Boyle's Law, **Equation #4A**, from Page 3-8, thus:

$$P_1 V_1 = P_2 V_2$$

Also we must remember that the volume of a sphere, in terms of its diameter is given by:

$$V_{sphere} = \frac{\pi d^3}{8}$$

We must now select the values that we will use for the "before" and "after" diameters of the balloon, as we try to develop a solution to this problem. Since the balloon is having its diameter decreased by 75%, we can select 4 units as the "before" diameter and 1 unit as the "after" diameter. Now applying Boyle's Law:

$$[1]\left[\frac{\pi[4]^3}{8}\right] = P_2 \left[\frac{\pi[1]^3}{8}\right] \quad \text{\& canceling, as appropriate:}$$

$$P_2 = 4^3 = 64$$

$\therefore P_2 = 64$ atmospheres

Problem #24:

This problem requires the use of Charles' Law, **Equation #4B**, from Page 3-9, thus:

$$\frac{V_1}{T_1} = \frac{V_2}{T_2}$$

The temperature difference with which we are dealing here is given by ΔT_F, and since STP conditions call for a temperature of $32°F = 0°C$, we can see that:

$$\Delta T_F = [-13°F - 32°F] = -45°F$$

Now converting this temperature [- 45°F] into degrees Celsius; thus:

$$\Delta T_C = \left\lfloor \frac{5}{9} \right\rfloor \Delta T_F$$

$$\Delta T_C = \left\lfloor \frac{5}{9} \right\rfloor [-45°F] = -25°C$$

Therefore, the temperature in the freezer is - 25°C, or converting into absolute temperature units, we have:

$$T_{freezer} = 273.16°K - 25°K = 248.16°K$$

We can now finally apply Charles' Law, after first converting the 1 liter volume of the balloon into its equivalent volume in milliliters [ie. 1 liter = 1,000 ml]; thus:

$$\frac{V_1}{T_1} = \frac{V_2}{T_2}$$

$$\frac{V_1}{248.16} = \frac{1,000}{273.16}$$

$$V_1 = \frac{248.16}{273.16}[1,000] = 908.5$$

Therefore, the final volume of the balloon is 908.5 ml; which means that its volume change — the quantity requested in the problem statement — is given by:

$$\Delta V = V_{final} - V_{initia}$$

$$\Delta V = 908.5 - 1,000 = -91.5$$

∴ The Change in the Balloon's Volume = - 92 ml.
(ie. Its Volume Decreased by 92 ml)

Problem #25:

Again, to solve this problem, we must apply Charles' Law, **Equation #4B**, from Page 3-9, thus:

$$\frac{V_1}{T_1} = \frac{V_2}{T_2}$$

$$\frac{46}{459.67°R + 77°R} = \frac{46.5}{T_2}$$

$$T_2 = \frac{[46.5][536.67]}{46} = \frac{24,955.16}{46} = 542.50$$

We must now convert this absolute temperature [in °R] back into degrees Fahrenheit, thus:

$$T_F = 542.50 - 459.67 = 82.83$$

∴ The Rupture Temperature of the Soap Bubble was 83°F

Problem #26:

This problem can be solved by using Gay-Lussac's Law, **Equation #4C**, which is on Page 3-9, thus:

$$\frac{P_1}{T_1} = \frac{P_2}{T_2}$$

$$\frac{950}{273.16°K + 3°K} = \frac{P_2}{273.16°K + 45°K}$$

$$P_2 = \frac{[950][318.16]}{276.16} = \frac{302,252}{276.16} = 1,094.5$$

∴ The Bottle's Eventual Internal Pressure Will Become 1,095 mm Hg at 45°C.

Problem #27:

This problem also can be solved by using Gay-Lussac's Law, **Equation #4C**, which is on **Page 3-9**, thus:

$$\frac{P_1}{T_1} = \frac{P_2}{T_2}$$

Note that the tire pressure in this problem is given in "psig", which means we will have to convert it into units absolute pressure, which for this case will be "psia"; thus:

$$P_{absolute} = P_{gauge} + 14.70$$

$$P_{absolute} = 32 + 14.70 = 46.70$$

Now, having the absolute pressure in "psia", we can finally apply Gay-Lussac's Law, thus:

$$\frac{P_1}{T_1} = \frac{P_2}{T_2}$$

$$\frac{46.7}{459.67°R + 56°R} = \frac{P_2}{459.67°R + 175°R}$$

$$P_2 = \frac{[46.7][634.67]}{515.67} = \frac{29,639.09}{515.67} = 57.48$$

We must now convert this absolute pressure, in "psia", back into its corresponding gauge pressure, in "psig", thus:

$$57.48 = P_{gauge} + 14.70$$

$$P_{gauge} = 57.48 - 14.70 = 42.78$$

∴ The Tire's Internal Pressure, When its Temperature Has Reached 175°F, will be 42 psig.

Problem #28:

This problem can be easily solved by using the General Gas Law, **Equation #28**, from Page 3-10, thus:

$$\frac{P_1 V_1}{T_1} = \frac{P_2 V_2}{T_2}$$

$$\frac{[5][1.0]}{459.67°R + 77°R} = \frac{[1]V_2}{459.67°R + 32°R}$$

$$V_2 = \frac{[5][491.67]}{536.67} = \frac{2,458.35}{536.67} = 4.58$$

∴ The Chlorine Resistant Balloon will have an STP Volume of 4.6 liters.

Problem #29:

This problem can also be easily solved by using the General Gas Law, **Equation #28**, from Page 3-10, thus:

$$\frac{P_1 V_1}{T_1} = \frac{P_2 V_2}{T_2}$$

To solve this problem, we must first calculate the total volume of N_2O that was delivered to the Operating Room. This can be done by multiplying the flow rate of N_2O by the time period during which the aforementioned flow rate occurred. It will also be helpful to convert the flow rate of 3,000 cc/min into a flow rate in liters per hour; thus:

$$V_{total} = \frac{[3,000][4.50][60]}{1,000} = 810 \text{ liters}$$

We must also convert the pressure of the cylinder of N_2O from psia to mm Hg, as follows:

$$P_{mm\ Hg} = \frac{[760]}{[14.70]} P_{psia}$$

$$P_{mm\ Hg} = \frac{[760][450]}{14.70} = \frac{342,000}{14.70} = 23,265.31$$

We can now finally apply the General Gas Law and obtain the desired result:

$$\frac{P_1 V_1}{T_1} = \frac{P_2 V_2}{T_2}$$

$$\frac{[23,265.31][30]}{T_1} = \frac{[766][810]}{273.16°K + 20°K}$$

$$T_1 = \frac{[23,265.31][30][293.16]}{[766][810]}$$

$$T_1 = \frac{204,613,714.3}{620,460} = 329.78$$

We must finally convert the resultant absolute temperature back to degrees Celsius, thus:

∴ The Sun Heated N_2O Cylinder Started out at a Temperature of 57°C = 134°F.

Problem #30:

This problem can be easily solved through the use of the Ideal Gas Law, **Equation #4E**, from Page 3-11, thus:

$$PV = nRT = \frac{m}{MW} RT$$

Now let us transpose this formula into a format that will provide a direct path to the answer that has been asked for in the statement of the problem; thus:

$$MW = \frac{mRT}{PV}$$

Here, to achieve consistency in the units, we shall use, as the Universal Gas Constant, the value:

R = **0.0821** [liter][atmospheres] per [°K][gram mole]

$$MW = \frac{[34][0.0821][273.16°K + 18°K]}{[0.92][8.58]}$$

$$MW = \frac{[34][0.0821][291.16]}{[0.92][8.58]} = \frac{812.74}{7.89} = 102.96$$

∴ Since the molecular weight of the refrigerant is approximately 102.96 amu, we can conclude that the material is most likely Freon 21, which has a molecular weight of 102.92 amu.

Problem #31:

This problem can also be easily solved by using the Ideal Gas Law, **Equation #4E**, from Page 3-11, thus:

$$PV = nRT = \frac{m}{MW}RT$$

Let us again transpose this formula into the same format as was used on the previous problem, a format that will provide a direct path to the answer that has been asked for in the statement of the problem, thus:

$$MW = \frac{mRT}{PV}$$

For this problem, as before with the previous problem, we must achieve consistency in the units we use, thus we shall employ for the Universal Gas Constant, the value:

R = **62.36** [liter][mm Hg] per [°K][gram mole]

$$MW = \frac{[5.00][62.36][273.16°K + 31°K]}{[755][4.36]}$$

$$MW = \frac{[5.00][62.36][304.16]}{[755][4.36]}$$

$$MW = \frac{94,837.09}{3,291.80} = 28.81$$

∴ The "Effective Molecular Weight" of Air is 28.81 amu.

Problem #32:

This problem will employ **Equation #5A**, from Page 3-12, thus:

$$n = \frac{W}{MW}$$

We must first determine the molecular weight [the formula weight] of $NaNO_3$:

$$MW_{NaNO_3} = [1][24.31] + [1][14.01] + [3][16.00]$$

$$MW_{NaNO_3} = 24.31 + 14.01 + 48.00 = 86.32$$

We can now apply **Equation #5A**, thus:

$$4.4 = \frac{W}{86.32}$$

$$W = [4.4][86.32] = 379.81$$

∴ 4.4 Moles of $NaNO_3$ Weighs 379.81 amu.

Problem #33:

To solve this problem, we must first apply **Equation #5B**, from Page 3-13, and then **Equation #5A**, from Page 3-12, thus:

$$n = \frac{Q}{N_A} = \frac{Q}{6.022 \times 10^{23}}$$

$$W_{grams} = \frac{1.045 \times 10^{-13}}{1,000,000,000} = 1.045 \times 10^{-22}$$

We know that Avogadro's Number [6.022×10^{23}] of the atoms of any element will constitute exactly 1.0 moles of that element; and we know that the numeric value of the weight, in grams, of 1.0 mole of any element will be equal, exactly, to the atomic weight of that element.

$$n = \frac{W}{MW}$$

$$1 = \frac{\left[6.022 \times 10^{23}\right]\left[1.045 \times 10^{-22}\right]}{MW}$$

$$MW = [6.022][1.045][10] = 62.93$$

∴ The Atomic Weight of this Isotope of Copper is 62.93 amu.

Problem #34:

To solve this problem, we must first apply **Equation #5C**, from Page 3-14, and then **Equation #5A**, from Page 3-12, thus:

$$n = \frac{V}{22.414}$$

We must now convert the volume of helium, given in cc, to liters — 3,400 cc = 3.4 liters. This will, of course, give us the first item requested in the problem statement.

$$n = \frac{3.4}{22.414} = 0.152$$

We can finally now apply **Equation #5A**, thus:

$$n = \frac{W}{MW}$$

$$0.152 = \frac{W}{4.00}$$

$$W = [0.152][4.00] = 0.607$$

∴ There are 0.152 Moles of Helium in 3,400 cc of this Gas at STP.
0.152 Moles of Helium Weigh 0.607 grams = 607 mg.

Problem #35:

To solve this problem, we must apply **Equation #5A**, from Page 3-12, and then deal with the basic definition for volume percent of a component, as this concept relates to the Mole Fraction of the same component.

Let us begin by assuming that we are dealing with <u>exactly</u> 100 grams of Oxyfume 12®. Obviously, this quantity of this material will contain 12 grams of ethylene oxide and 88 grams of Freon 12. We must next determine the number of moles of each of these materials that would be in this 100 gram quantity; this will give us the necessary information to determine the Mole Fractions of each, thus:

$$n = \frac{W}{MW}$$

1. For ethylene oxide:
$$n_{ethylene\ oxide} = \frac{12}{44.05} = 0.272$$

2. For Freon 12:
$$n_{Freon\ 12} = \frac{88}{120.92} = 0.728$$

We must next determine the Mole Fraction of each of the components in this mixture; since by definition, the Mole Fraction for each component can be converted to its volume percentage simply by multiplying by 100.

$$n_{total} = 0.272 + 0.728 = 1.000$$

1. For ethylene oxide:
$$Mole\ Fraction_{ethylene\ oxide} = \frac{0.272}{1.000} = 0.272$$

2. For Freon 12:
$$Mole\ Fraction_{Freon\ 12} = \frac{0.728}{1.000} = 0.728$$

∴ Oxyfume 12® is made up, as follows:
ethylene oxide: 27.2% by volume
Freon 12: 72.8% by volume

Problem #36:

To solve this problem, we must apply **Equation #6A**, from Page 3-15, thus:

$$\frac{\rho_1 T_1}{P_1} = \frac{\rho_2 T_2}{P_2}$$

As with several of the earlier problems, we must initially determine the correct absolute temperature, in °R, thus:

$$T_1 = 459.67°R + 55°R = 514.67°R$$

Also, we must remember the relationship, as stated in the problem, between the starting density of the cool, unheated air, and the density of this same air, once it has been heated to a sufficiently high temperature to achieve a liftoff of the 1,650 lbs load, thus:

$$\frac{\rho_{hot\ air}}{\rho_{cool\ air}} = 0.95 \quad \text{and, therefore:}$$

$$\rho_{hot\ air} = [0.95]\rho_{cool\ air}$$

At this point, we can return to using **Equation #6A** to develop a solution to the problem, thus:

$$\frac{\rho_1 T_1}{P_1} = \frac{\rho_2 T_2}{P_2}$$

Note, finally that the atmospheric pressure, designated below as "P", both before and after the balloon liftoff are the same, and that this is the pressure that characterizes both the heated air in the balloon envelope, and the unheated ambient air that surrounds it. As such, it and the density of the cool air will cancel; thus:

$$\frac{[0.95][\rho_{cool\ air}]T_1}{P} = \frac{[\rho_{cool\ air}][514.67]}{P}$$

$$T_1 = \frac{514.67}{0.95} = 541.76$$

Finally, we must determine the change in temperature that was necessary to achieve the liftoff, thus:

$$\Delta T = 541.76 - 514.67 = 27.09$$

> ∴ The air in the balloon envelope must undergo an increase in temperature of 27.1°F in order to achieve a liftoff of Phineas Fogg's hot air balloon.

Problem #37:

To solve this problem, we must again use **Equation #6A**, from Page 3-15, thus:

$$\frac{\rho_1 T_1}{P_1} = \frac{\rho_2 T_2}{P_2}$$

Here we are dealing with air at two clearly different conditions, and the solution can be obtained simply by substituting directly into this formula and solving. We must first convert all the temperatures from the listed °F into an appropriate absolute temperature scale (ie. °R), thus:

$$\frac{[1,400°R + 459.67°R]\rho_1}{1400} = \frac{[459.67][1.182]}{760}$$

$$\rho_1 = \frac{[459.67][1.182][1,400]}{[760][1,859.67]}$$

$$\rho_1 = \frac{460,661.92}{1,413,349.20} = 0.538$$

> ∴ The Density of Air at the Booster Inlet will be 0.538 mg/cm³.

Problem #38:

To solve this problem, we must use Dalton's Law of Partial Pressures, **Equation #7B**, from Page 3-17. Remember that the volume percentage of any component in a gas mixture, when divided by 100, will give the Mole Fraction of that component; thus:

$$P_i = P_{total}\, m_i$$

1. For oxygen: $P_{oxygen} = [760][0.209] = 158.84$

$$P_{oxygen} = 158.8 \text{ mm Hg}$$

2. For argon:
$$P_{argon} = [760][0.009] = 6.84$$

$$P_{argon} = 6.8 \text{ mm Hg}$$

∴ The Partial Pressures of these Two Components of Air under NTP Conditions are:

the Partial Pressure of Oxygen = 158.8 mm Hg
the Partial Pressure of Argon = 6.8 mm Hg

Problem #39:

To solve this problem, we must again use Dalton's Law of Partial Pressures, **Equation #7B**, from Page 3-17. For this problem, too, recall that the volume percentage of any component in a gas mixture, when divided by 100, will give the Mole Fraction of that component, thus:

$$P_i = P_{total} \, m_i$$

1. For the LEL Condition:
$$P_{LEL} = [1,013.25][0.021] = 21.28$$

$$P_{LEL} = 21.3 \text{ millibars}$$

2. For the UEL Condition:
$$P_{UEL} = [1,013.25][0.095] = 96.26$$

$$P_{UEL} = 96.3 \text{ millibars}$$

∴ The Partial Pressure of Propane at its LEL = 21.3 millibars
& its Partial Pressure at its UEL = 96.3 millibars.

Problem #40:

To solve this problem, we must use Raoult's Law, **Equation #7C**, from Page 3-18, and then **Equation #7B**, from Page 3-17. We must first determine the Mole Fractions of ethanol in Irish Whiskey, by analyzing a 100 gram sample of this material, which sample will obviously contain 45 grams of ethanol and 55 grams of water, thus:

1. For ethanol:
$$n_{ethanol} = \frac{45}{46.07} = 0.977$$

2. For water:
$$n_{water} = \frac{55}{18.02} = 3.052$$

Now that we know the number of moles of each of the two components of Irish Whiskey, we can determine the Mole Fraction of the ethanol component; thus:

$$m_{ethanol} = \frac{0.977}{0.977 + 3.052} = \frac{0.977}{4.029} = 0.242$$

Since we now know the Mole Fraction of this material, we can determine its partial vapor pressure through the use of Raoult's Law; thus:

$$PVP_i = m_i VP_i$$

$$PVP_{ethanol} = [0.242][44] = 10.67$$

It is now easy to apply **Equation #7B** to obtain the equilibrium concentration of ethanol in this room, thus:

$$PVP_i = \frac{P_{total} C_i}{1,000,000}$$

Transposing **Equation #7B** into a more useful form for the purposes of this problem, we have:

$$C_i = \frac{[1,000,000][PVP_i]}{P_{total}}$$

Finally, substituting in the appropriate values that have just been determined, or that were given in the problem statement, we have:

$$C_{ethanol} = \frac{[1,000,000][10.67]}{765}$$

$$C_{ethanol} = \frac{10,670,000}{765} = 13,947.7$$

∴ The ultimate, ambient, evaporative equilibrium concentration of ethanol that will be achieved in this room = 13,950 ppm.

Problem #41:

To solve this problem, we will have to employ **Equation #7B**, on Page 3-17, then **Equation #7C**, on Page 3-18. We must start by determining the partial pressure of ethanol if it is at a concentration of 1,000 ppm, using **Equation #7B**; thus:

$$PVP_i = \frac{P_{total}\, C_i}{1,000,000}$$

$$PVP_{ethanol} = \frac{[765][1,000]}{1,000,000} = \frac{765,000}{1,000,000} = 0.765$$

We can now apply **Equation #7C** to obtain the Mole Fraction of ethanol that must exist in the liquid phase in order for the partial vapor pressure to be as calculated above, thus:

$$PVP_i = m_i VP_i$$

Transposing into a more useful format for determining the Mole Fraction of ethanol, we have:

$$m_{ethanol} = \frac{PVP_{ethanol}}{VP_{ethanol}}$$

$$m_{ethanol} = \frac{0.765}{44} = 0.0174$$

Now, we must determine the total number of moles of ethanol that were in the original shot glass of Irish Whiskey. We have been told that the specific gravity of Irish Whiskey is the same as that of water (ie. its density = 1.00 grams/cm³); and we know that this glass started with 200 ml of this liquid. We can, therefore, conclude that there was a total of 90 grams of Irish Whiskey and 110 grams of water in the shot glass to start off with. We must now determine the starting number of moles of each of the two components in the shot glass.

1. For ethanol:

$$n_{ethanol} = \frac{W_{ethanol}}{MW_{ethanol}} = \frac{90}{46.07} = 1.954$$

2. For water:

$$n_{water} = \frac{W_{water}}{MW_{water}} = \frac{110}{18.02} = 6.104$$

Since we know the ultimate Mole Fraction of ethanol that is required to achieve the target ambient ethanol concentration, we can apply the following Equation to determine the amount of water that must be added to achieve the desired result, thus:

$$m_{ethanol} = \frac{n_{ethanol}}{n_{water} + n_{water-added} + n_{ethanol}}$$

$$0.0174 = \frac{1.954}{6.104 + n_{water-added} + 1.954} = \frac{1.954}{n_{water-added} + 8.058}$$

$$n_{water-added} + 8.058 = \frac{1.954}{0.0174} = 112.273$$

$$n_{water-added} = 112.273 - 8.058 = 104.215$$

Now that we know the number of moles of water that must be added, we can readily convert this to a volume of water, which is what was asked for in the problem statement. We do this by first determining the weight of water that must be added; and then, since the density of water is 1.000 grams/cm^3, we see that the number of grams added and the volume added (when measured in cm^3), are numerically the same! Thus:

$$W_{water-added} = [104.215][18.02] = 1,877.95$$

∴ The volume of water that must be added to this shot glass in order to achieve the desired final ambient concentration of ethanol is 1,878 ml, or 1.89 liters! Clearly, we will have to transfer the contents of the shot glass to a fairly large pitcher in order to accommodate the additional water that will be required.

Problem #42:

To solve this problem, we must use **Equation #8A**, from Page 3-19, thus:

$$V_{s-cm/hr} = 10.797 \rho d^2$$

$$V_{s-cm/hr} = [10.797][1.848][1.5]^2$$

$$V_{s-cm/hr} = [10.797][1.848][2.25]$$

$$V_{s-cm/hr} = 44.89$$

∴ It is likely that the Industrial Hygienist calculated a Settling Velocity of 44.9 cm/hour (or 17.7 inches/hour) for the beryllium particulates on the stamping floor of this Metallic Spring Manufacturing Company.

Problem #43:

To solve this problem, we must again use **Equation #8A**, from Page 3-19, thus:

$$V_{s-cm/hr} = 10.797\rho d^2$$

To make the solution to this problem simpler, let us convert this relationship so that it provides a Settling Velocity in inches/hour, rather than cm/hour. To do this, we must divide by 2.54, thus:

$$V_{s-in/hr} = 4.251\rho d^2$$

$$72 = [4.251][1.848]d^2 \quad \text{and solving:}$$

$$d^2 = \frac{72}{[4.251][1.848]} = \frac{72}{7.86} = 9.17$$

$$d = \sqrt{9.17} = 3.03$$

Since the initial average diameter of the beryllium particles was determined to have been 1.5 microns, and since this diameter must now be increased to 3.03 microns, in order to achieve the desired Settling Velocity, we see that we must increase the particle diameter, quantitatively by the following diameter multiplier, "DM":

$$DM = \frac{3.03}{1.5} = 2.02$$

∴ It appears as if the Industrial Hygienist will be forced to recommend the "C" category of Silicone Coating. The "B" category material will almost work; however, if the requirement truly is for a minimum of 72 inches/hour Settling Velocity, then the recommendation will have to be for the "C" material.

Problem #44:

To solve this problem, we must use **Equation #9A**, from Page 3-20; however, we must first convert the wavelength (given in microns) to centimeters, thus:

$$f = \frac{c}{\lambda}$$

$$f_{\text{C-H Stretch}} = \frac{3 \times 10^{10}}{3.35 \times 10^{-4}} = 8.95 \times 10^{13}$$

∴ The frequency of the infrared photon at which the carbon-hydrogen bond absorbs is 8.95×10^{13} hertz = 8.95×10^{13} cycles/second = 8.95×10^{7} megahertz.

Problem #45:

To solve this problem, we must again use **Equation #9A**, from Page 3-20, prior to applying this formula, we must first convert the frequency (given in megahertz) into hertz, thus:

$$\lambda = \frac{c}{f}$$

$$\lambda = \frac{3 \times 10^{10}}{2.84 \times 10^{20}} = 1.06 \times 10^{-10}$$

∴ the desired wavelength is 1.06×10^{-10} centimeters, or 1.06×10^{-6} microns.

Problem #46:

To solve this problem, we must again use **Equation #9B**, from Page 3-21, thus:

$$\nu = \frac{1}{\lambda}$$

1. For the carbon-hydrogen stretch photon:

$$\nu_{\text{C-H Stretch}} = \frac{1}{3.35 \times 10^{-4}}$$

$$\nu_{\text{C-H Stretch}} = 2,985$$

2. For the Cobalt-60 gamma ray photon:

$$\nu_{\text{Co - }\gamma\text{ Ray}} = \frac{1}{1.056 \times 10^{-10}}$$

$$\nu_{\text{Co - }\gamma\text{ Ray}} = 9.47 \times 10^{9}$$

$\therefore$ The two wavenumbers requested are:
The carbon-hydrogen stretch photon: $\nu = 2,985$ cm^{-1}
The Cobalt-60 gamma ray photon: $\nu = 9.47 \times 10^{9}$ cm^{-1}

Problem #47:

To solve this problem, we must use **Equation #10A**, from **Page 3-22**, thus:

$$N_t = N_0 e^{-kt}$$

1. For January 20th (19 days):

$$N_{t_1} = \left[2.0 \times 10^{-6} \right] e^{-[0.0862][19]}$$

$$N_{t_1} = \left[2.0 \times 10^{-6} \right] e^{-1.638}$$

$$N_{t_1} = \left[2.0 \times 10^{-6} \right] [0.194] = 3.88 \times 10^{-7}$$

2. For One Year Later (365 days):

$$N_{t_2} = \left[2.0 \times 10^{-6} \right] e^{-[0.0862][365]}$$

$$N_{t_2} = \left[2.0 \times 10^{-6} \right] e^{-31.463}$$

$$N_{t_2} = \left[2.0 \times 10^{-6} \right] \left[2.17 \times 10^{-14} \right]$$

$$N_{t_2} = 4.33 \times 10^{-20}$$

$\therefore$ On January 20th, there will be only 0.39 µg of this isotope remaining (19.4% of the material that was received on January 1st). One year later, there will be only 4.33 x 10^{-20} µg of this isotope remaining — probably an undetectable amount.

Problem #48:

To solve this problem, we must again use **Equation #10A**, from Page 3-22, thus:

$$N_t = N_0 e^{-kt}$$

$$1.5 \times 10^{-10} = 8.8 \times 10^{-6} e^{-[0.502]t}$$

$$e^{-[0.502]t} = \frac{1.5 \times 10^{-10}}{8.8 \times 10^{-6}} = 0.000017$$

$$-[0.502]t = -10.98$$

$$t = \frac{-10.98}{-0.502} = 21.87$$

∴ This Scientist must use her supply of $_{99}Es^{245}$ in the next 21.87 ~ 22 minutes.

Problem #49:

To solve this problem, we must use **Equation #10B** from Page 3-23, thus:

$$k = \frac{0.693}{T_{1/2}}$$

$$k = \frac{0.693}{12.26} = 0.0565$$

∴ The Radioactive Decay Constant for $_1H^3$ is: $k_{tritium} = 0.0565$ years^{-1}

Problem #50:

The solution to this problem also requires the use of **Equation #10B** from Page 3-23, thus:

$$T_{1/2} = \frac{0.693}{k}$$

1. For $_{53}I^{131}$:

$$T_{1/2} = \frac{0.693}{0.0862} = 8.039$$

$$T_{1/2} = 8.04 \text{ days}$$

2. For $_{99}Es^{245}$:

$$T_{1/2} = \frac{0.693}{0.502} = 1.380$$

$$T_{1/2} = 1.38 \text{ minutes}$$

∴ The requested Half Lives are as follows:

For Iodine [$_{53}I^{131}$]: $T_{1/2} = 8.04$ days

For Einsteinium [$_{99}Es^{245}$]: $T_{1/2} = 1.38$ minutes

Problem #51:

The solution to this problem will require the use of **Equation #10C** from **Page 3-24**, thus:

$$A_t = \frac{0.693}{T_{1/2}} N_0 e^{-\frac{0.693\, t}{T_{1/2}}}$$

Before directly addressing the solution to this problem by using the foregoing Equation, we must first convert the Half Life of Americium from the form in which it has been given in the problem statement (that being in units of "years") to units of "seconds", since the problem is to determine the number of disintegrations per sec-ond; thus:

$$T_{1/2} = [432.2][365][24][60][60] = 1.363 \times 10^{10} \text{ seconds}$$

We are interested in the specific number of disintegrations that are occurring in each second, at that instant in time when the actual mass of Americium present is exactly 1.75 µg; thus as we substitute numbers and values into **Equation #10C**, we shall use a time, t = 0 seconds, thus:

$$A_t = \frac{0.693}{1.363 \times 10^{10}} [1.75]\left[4.0 \times 10^{15}\right] e^{-\frac{[0.693][0]}{1.363 \times 10^{10}}}$$

$$A_t = \frac{[0.693][1.75][4.0 \times 10^{15}]}{1.363 \times 10^{10}} e^0$$

$$A_t = \frac{[0.693][1.75][4.0 \times 10^{15}]}{1.363 \times 10^{10}}$$

$$A_t = \frac{4.851 \times 10^{15}}{1.363 \times 10^{10}} = 355,910$$

∴ A Smoke Detector will only operate properly if there are 355,910 (or more) Americium atoms disintegrating, or decaying, each second.

Problem #52:

To solve this problem, we must first use **Equation #10B** from Page 3-23 to determine the Radioactive Decay Constant of Americium, and then **Equation #10A** from Page 3-22 to develop the answer that has been requested, thus:

$$k = \frac{0.693}{T_{1/2}}$$

$$k = \frac{0.693}{432.2} = 0.00160 \text{ years}^{-1}$$

Now with this Radioactive Decay Constant, we can apply **Equation #10A**, thus:

$$N_t = N_0 e^{-kt}$$

$$1.75 = 1.80 e^{-[0.00160]t}$$

$$\frac{1.75}{1.80} = e^{-[0.00160]t} = 0.9722$$

$$ln[0.9722] = ln\left[e^{-[0.00160]t}\right]$$

$$-0.0282 = -0.00160t$$

$$t = \frac{-0.0282}{-0.00160} = 17.61$$

∴ It will take 17.6 years for the amount of Americium provided in a new Smoke Detector to decay to the extent that there will be fewer than the minimum required 355,910 disintegrations/second occurring. It is unlikely that we can consider that the manufacturer has "built in" product obsolescence as a result of this "lifetime", since this time period (17.6 years) is simply too long. I cannot imagine that the average Smoke Detector owner would note that he or she will have to purchase a new unit 17+ years after the date of his or her current purchase; rather, after 17+ years, that person's Smoke Detector would simply cease to function properly without his or her knowing it!

Problem #53:

To solve this problem, we must employ **Equation #11A** from Page 3-25, thus:

$$E = \frac{\Gamma A}{d^2}$$

$$E = \frac{[18.81][550]}{[100]^2} = \frac{10,345.50}{10,000} = 1.035$$

∴ The Requested Dose Exposure Rate = 1.04 Rads/hour.

Problem #54:

The solution to this problem also employs **Equation #11A** from Page 3-25, thus:

$$E = \frac{\Gamma A}{d^2}$$

$$48.0 = \frac{440\Gamma}{[30]^2}$$

$$\Gamma = \frac{[48.0][30]^2}{440}$$

$$\Gamma = \frac{43,200}{440} = 98.18$$

∴ The Radiation Constant for $_{88}Ra^{226}$ is 98.2 Rad•cm^2/hr•mCi.

Problem #55:

To solve this problem, we must use **Equation #55** on Page 3-26, thus:

$$D_{Rem} = D_{Rad}[QF]$$

For γ-rays, the Quality Factor = QF = 1.0, thus:

   1. For the closed aperture time intervals: $D_{Rem_1} = [0.09][5.8][1.0]$

$$D_{Rem_1} = 0.522$$

   2. For the open aperture time intervals: $D_{Rem_2} = [0.44][2.2][1.0]$

$$D_{Rem_2} = 0.968$$

To obtain the Total Dose to which this Radiation Technician is exposed on a daily basis, we must employ the following:

$$D_{Rem\ Total} = \sum_{i=1}^{2} D_{Rem\ i} = D_{Rem_1} + D_{Rem_2}$$

$$D_{Rem\ Total} = 0.522 + 0.968 = 1.49$$

∴ This Radiation Technician's adjusted Radiation Dose = 1.49 Rem/day.

Problem #56:

The solution to this problem also requires the use of **Equation #11B** from Page 3-26, thus:

$$D_{Sieverts} = D_{Grays}[QF]$$

For Thermal Neutrons, the Quality Factor = QF = 5.0, and — for the Technicians involved in this situation — we are dealing with a 5 day workweek, with each day consisting of 8 hours, thus:

1. For the closed access port time intervals:

$$D_{\text{Sieverts}_1} = [0.12][38][5.0]$$

$$D_{\text{Sieverts}_1} = 22.80$$

2. For the open access port time intervals:

$$D_{\text{Sieverts}_2} = [0.83][2][5.0]$$

$$D_{\text{Sieverts}_2} = 8.30$$

To obtain the Total Dose to which each of these Technicians are exposed on a weekly basis, we must employ the following:

$$D_{\text{Sieverts Total}} = \sum_{i=1}^{2} D_{\text{Sieverts}_i} = D_{\text{Sieverts}_1} + D_{\text{Sieverts}_2}$$

$$D_{\text{Sieverts Total}} = 22.80 + 8.30 = 31.10$$

∴ These Technicians have an Accumulated Dose Rate of 31.1 Sieverts/week.

Problem #57:

To solve this problem, we must use **Equation #12A** from Page 3-27, thus:

$$ER_{\text{goal}} = \frac{ER_{\text{source}}}{2^{x/\text{HVL}}}$$

We must consider two cases, namely: (1) a shielded cell made of concrete; and (2) a shielded cell made of lead; we shall assume *relative* target "effective" Radiation Emission Rates for these two situations — ie. **775** for the concrete shielded cell case, and **1.0** for the lead shielded cell case; thus:

1. For a concrete shielded cell:

$$ER_{\text{goal}_{\text{concrete}}} = \frac{ER_{\text{source}}}{2^{x_{\text{concrete}}/\text{HVL}_{\text{concrete}}}} \quad \text{\& substituting:}$$

$$775 = \frac{ER_{\text{source}}}{2^{18/2.45}} = \frac{ER_{\text{source}}}{2^{7.35}} = \frac{ER_{\text{source}}}{162.80}$$

2. For a lead shielded cell: $\quad ER_{goal\,Pb} = \dfrac{ER_{source}}{2^{x_{Pb}/HVL_{Pb}}}$ & substituting:

$$1 = \dfrac{ER_{source}}{2^{8/HVL_{Pb}}}$$

Now, dividing the first derived equation by the second, we get a relationship where the ER_{source}'s cancel out, leaving as the only unknown the required HVL_{Pb}; thus:

$$\frac{775}{1} = \frac{\dfrac{ER_{source}}{162.80}}{\dfrac{ER_{source}}{2^{8/HVL_{Pb}}}} \quad \text{& simplifying:}$$

$$775 = \frac{2^{8/HVL_{Pb}}}{162.80}$$

$[775][162.80] = 2^{8/HVL_{Pb}} = 126,168.42$ & taking the logarithm of both sides:

$$log\,126,168.42 = [log\,2]\left[\frac{8}{HVL_{Pb}}\right]$$

$$5.101 = \frac{[0.301][8]}{HVL_{Pb}}$$

$$HVL_{Pb} = \frac{[0.301][8]}{5.101} = 0.472$$

∴ Thus, we see that the Half Value Thickness of Lead is 0.47 inches = 1.20 cm.

Problem #58:

The solution to this problem requires the use of **Equation #12B** from Page 3-28, thus:

$$\frac{ER_A}{ER_B} = \frac{S_B^2}{S_A^2}$$

$$\frac{ER_A}{200} = \frac{[15]^2}{[25]^2} = \frac{225}{625} = 0.36$$

$$ER_A = [200][0.36] = 72$$

∴ The Doctor would experience a Radiation Intensity of 72 Sieverts/hour.

Problem #59:

The solution to this problem will require, initially, the use of the Quality Factor Table associated with **Equation #11B** from Page 3-26, followed by the use of **Equation #12B** from Page 3-28. We must first determine what impact on the adjusted Radiation Dose, in mRem, would be as a result of the change in the character of the radiation (ie. going from an Alpha emitter to a β-emitter) thus:

$$\frac{QF_{\beta-Rays}}{QF_{Alpha\ Rays}} = \frac{1}{20}$$

Thus we see that simply changing the source radiation type will diminish the experienced Radiation Dose by a factor of 20.

∴ The adjusted Radiation Dose of the $_{19}K^{45}$ source, vs. the $_{88}Ra^{266}$ source would be: 2 mRem/hour [$_{19}K^{45}$]vs. the stated 40 mRem/hr [$_{88}Ra^{266}$].

Now we can apply **Equation #12B** from Page 3-28 to obtain the final requested answer; thus:

$$\frac{ER_A}{ER_B} = \frac{S_B^2}{S_A^2}$$

$$\frac{2}{40} = \frac{S_B^2}{[25]^2}$$

$$S_B^2 = \frac{[2][25]^2}{40} = \frac{[2][625]}{40}$$

$$S_B^2 = 31.25$$

$$\sqrt{S_B^2} = \sqrt{31.25}$$

$$S_B = 5.59$$

∴ To obtain the same 40 mRem/hour adjusted Dose Rate, the receptor would have to be located 5.6 cm from the source.

Problem #60:

To solve this problem, we must use **Equation #13A** from Page 3-29, thus:

$$OD = log \left| \frac{I_{incident}}{I_{transmitted}} \right|$$

$$OD = log \left| \frac{475}{.45} \right| = log[1,055.56]$$

$$OD = 3.02$$

∴ The Optical Density of the Operator's Goggles = OD = 3.02.

Problem #61:

This problem also requires the use of **Equation #13A** from Page 3-29, thus:

$$OD = log \left| \frac{I_{incident}}{I_{transmitted}} \right|$$

$$OD = log \left| \frac{[475][6.62]}{0.19} \right|$$

$$OD = log \left| \frac{3,144.5}{0.19} \right| = log[16,550]$$

$$OD = 4.22$$

We have not been asked simply for the new Optical Density, rather we have been asked for the increase in Optical Density under the new conditions; thus we shall proceed as follows:

$$\Delta OD = \frac{OD_{new}}{OD_{former}}$$

$$\Delta OD = \frac{4.22}{3.02} = 1.40$$

∴ The Optical Density of the Goggles will have to be increased by 140%.

Problem #62:

This problem requires the use of **Equation #14A** from Page 3-30, thus:

$$r_{FF} = \frac{\pi D^2}{8\lambda}$$

Remembering, also, that the antenna diameter term, D, in this Equation must be expressed in "centimeters", rather than in the "inches" given in the Problem Statement, we must convert the latter to the former by multiplying by 2.54 cm/inch.

$$r_{FF} = \frac{[\pi][40.5]^2[2.54]^2}{[8][46]}$$

$$r_{FF} = \frac{33,245.08}{368} = 90.34$$

∴ The Far Field is 90.3 cm = 35.6 inches out in front of this UHF antenna.

Problem #63:

This problem can be solved by the application of **Equation #14B** from Page 3-31.

In addition, the well known relationship for determining TWAs will have to be used to obtain the final part of the answer requested in this problem. It should be noted that we are dealing here in the Near Field; since, from the previous problem, we determined that the Far Field started at a distance of 35.6 inches out in front of this antenna; thus:

$$W_{NF} = \frac{16P}{\pi D^2}$$

There appear to be two unit conversions that must be made in the factors that are to be used in this Equation. First, we have been given the antenna output power in "kilowatts"; however, the Equation requires that the power term, P, be expressed in "milliwatts". To obtain the latter from the former we will have to multiply by the factor, 10^6 mw/kw. Also, we again have the antenna diameter in "inches", but must have it in "centimeters"; thus we must again multiply by 2.54 cm/inch.

$$W_{NF} = \frac{[16][0.05][10]^6}{\pi[40.5]^2[2.54]^2}$$

$$W_{NF} = \frac{800,000}{33,245.08} = 24.06$$

∴ The Power Density 12 inches in front of this UHF antenna is 24.06 mw/cm^2.

We must next determine the time interval during which the Service Technician can safely work in this position. We shall employ the standard Time Weighted Average formula; thus:

$$TWA = \frac{\sum_{i=1}^{n} T_i P_i}{\sum_{i=1}^{n} T_i} = \frac{T_1 P_1 + T_2 P_2 + \cdots + T_n P_n}{T_1 + T_2 + \cdots + T_n}$$

We shall assume that this Service Technician will spend "T" minutes (out of every 6 minute time interval) working at a point that is 12 inches in front of this antenna, and the remaining time, "6 - T" minutes, well away from this antenna — in a location where the Power Density = 0.0 mw/cm^2. We shall further assume that this Service Technician's 6-minute TWA was at the TLV-TWA level of 6 mw/cm^2; thus:

$$6 = \frac{24.06 T + 0.0[6 - T]}{T + [6 - T]} = \frac{24.06 T}{6}$$

$$24.06 T = [6][6] = 36$$

$$T = \frac{36}{24.06} = 1.50$$

∴ This Service Technician will only be able to work for a period of 1.5 minutes (1 minute, 30 seconds), or less, at a point 12 inches in front of this antenna, when it is transmitting. If he works for a longer period of time, he will have exceeded the TLV-TWA, which is 6 mw/cm^2 .

Problem #64:

This problem will require the use of **Equation #14D** from Page 3-32, thus:

$$r = \frac{D}{2\lambda} \sqrt{\frac{\pi P}{W_{FF}}}$$

$$r = \frac{[40.5][2.54]}{[2][46]} \sqrt{\frac{\pi[0.05][10]^6}{6}}$$

$$r = \frac{102.87}{92} \sqrt{\frac{157,079.63}{6}} = 1.118\sqrt{26,179.94}$$

$$r = [1.118][161.80] = 180.92$$

Although we did not know if this Power Density Level would be in the Far Field (in contrast to the Near Field) at the start of this problem, we can now see that it truly is in the Far Field (which starts 90.3 cm from the antenna). Had this value been less than 90.3 cm, we would have had to use **Equation #14B** from Page 3-31.

∴ Although it is difficult to determine how it would actually be done at the calculated distance, the closest that a Service Technician could safely work in front of this UHF antenna (and be certain <u>never</u> to exceed the 6-minute TLV-TWA of 6 mw/cm^2) would be at a distance of 180.9 cm = 71.2 inches.

Problem #65:

This problem will also require the use of **Equation #14D** from Page 3-32. Clearly, this point is in the Far Field. In fact, we will have to convert this very large distance into the units required for this Equation; thus:

$$r = \frac{D}{2\lambda} \sqrt{\frac{\pi P}{W_{FF}}}$$

Remembering, again, that the antenna diameter term, D, in this Equation must be expressed in "centimeters", rather than in the "inches" given in the Problem Statement, we must convert the latter to the former by multiplying by 2.54 cm/inch. We must also convert the output power term, P, from the "kilowatts" in which it has been given in the Problem Statement to the "milliwatts" that are required by the Equation — thus we must again multiply by 10^6 mw/kw.

$$[85][5,280][12][2.54] = \frac{[40.5][2.54]}{[2][46]} \sqrt{\frac{\pi[0.05][10]^6}{W_{FF}}}$$

$$1.368 \times 10^7 = \frac{102.87}{92} \sqrt{\frac{157,079.63}{W_{FF}}}$$

$$\frac{[1.368 \times 10^7][92]}{102.87} = \sqrt{\frac{157,079.63}{W_{FF}}} = 12,233,955.56$$

$$\frac{157,079.63}{W_{FF}} = 1.497 \times 10^{14}$$

$$W_{FF} = \frac{157,079.63}{1.497 \times 10^{14}} = 1.050 \times 10^{-9}$$

∴ The minimum Power Density required for this UHF receiving antenna to be able to operate successfully appears to be 1.05 x 10⁻⁹ mw/cm² = 1.05 x 10⁻⁶ µw/cm².

Problem #66:

This problem must be solved by first **Equation #9A** from Page 3-20, to determine the wavelength of this X-Band radar microwave, and then using **Equation #14A** from Page 3-30; thus:

$$\lambda = \frac{c}{f}$$

$$\lambda = \frac{3 \times 10^{10}}{10.525 \times 10^9} = 2.85$$

Now since we know the wavelength of this X-Band radar microwave, we can now determine the distance to the Far Field, using **Equation #14A**; thus:

$$r_{FF} = \frac{\pi D^2}{8\lambda}$$

$$r_{FF} = \frac{\pi[4.02]^2}{[8][2.85]} = 2.23$$

∴ It is only 2.23 cm = 0.88 inches to the Far Field for this X-Band Speed Radar Gun.

Problem #67:

This problem likely will use **Equation #14D** from Page 3-32 to obtain a solution. In the event that the distance we calculate turns out to be less than 2.24 cm, then we will have to recalculate using **Equation #14B** from Page 3-31; thus:

$$r = \frac{D}{2\lambda}\sqrt{\frac{\pi P}{W_{FF}}}$$

$$r = \frac{[4.02][2.54]}{[2][2.85]}\sqrt{\frac{\pi[45]}{10}}$$

$$r = \frac{10.21}{5.70}\sqrt{\frac{141.37}{10}} = 1.79\sqrt{14.137}$$

$$r = [1.79][3.76] = 6.74$$

∴ The distance in front of this X-Band Speed Radar Gun at which the Power Density Level will be at the 10 mw/cm^2 (or less) will be 6.74 cm = 2.65 inches (or greater). For any distance greater than this, there will <u>never</u> be any danger of an individual exceeding the established TLV-TWA of 10 mw/cm^2; however, at distances less than this value, there will be a potential problem with exceeding this Standard.

Problem #68:

To solve this problem we will have to employ **Equation #15B,** from Page 3-34, since this problem involves evaluating the WGBT Index <u>with</u> a solar load (subscript = Outside); thus:

$$WBGT_{Outside} = 0.7[NWB] + 0.2[GT] + 0.1[DB]$$

$$WBGT_{Outside} = [0.7][72] + [0.2][102] + [0.1][88]$$

$$WBGT_{Outside} = 50.40 + 20.40 + 8.8 = 79.6$$

$$\therefore \text{The } WBGT_{Outside} \text{ Index} = 79.6°F = 26.4°C$$

Problem #69:

To solve this problem we will have to employ **Equation #15A,** from Page 3-33, since this problem involves evaluating the WGBT Index <u>without</u> a solar load (sub-script = Inside). An additional subtlety to this problem is the seeming lack of a Wet Bulb Temperature. The clue to the answer, however, is in the problem statement. When it is raining, the relative humidity is 100%, and the wet bulb temperature is equal to the dry bulb temperature; thus:

$$WBGT_{Inside} = 0.7[NWB] + 0.3[GT]$$

$$WBGT_{Inside} = [0.7][78] + [0.3][102]$$

$$WBGT_{Inside} = 54.6 + 30.6 = 85.2$$

$$\therefore \text{The } WBGT_{Inside} \text{ Index} = 85.2°F = 29.6°C.$$

Problem #70:

To solve this problem, we will have to employ **Equation #15A,** from Page 3-33, first, and then **Equation #15A,** from Page 3-35; thus:

$$\text{WBGT}_{\text{Inside}} = 0.7[\text{NWB}] + 0.3[\text{GT}]$$

1. For the Open Hearth Area:

$$\text{WBGT}_{\text{Inside}_1} = 0.7[117] + 0.3[175]$$

$$\text{WBGT}_{\text{Inside}_1} = 81.9 + 52.5 = 134.4$$

2. For the Operator Rest Area:

$$\text{WBGT}_{\text{Inside}_2} = 0.7[63] + 0.3[75]$$

$$\text{WBGT}_{\text{Inside}_2} = 44.1 + 22.5 = 66.6$$

3. For Elsewhere in the Mill:

$$\text{WBGT}_{\text{Inside}_3} = 0.7[102] + 0.3[93]$$

$$\text{WBGT}_{\text{Inside}_3} = 71.4 + 27.9 = 99.3$$

With these three WBGT$_{\text{Inside}}$ Indices known, we can now employ **Equation #16A** from Page 3-35, to obtain the final TWA value requested; thus:

$$\text{WBGT}_{\text{TWA}} = \frac{\sum_{i=1}^{n} \text{WBGT}_i T_i}{\sum_{i=1}^{n} T_i} = \frac{\sum_{i=1}^{n} \text{WBGT}_1 T_1 + \text{WBGT}_2 T_2 + \cdots + \text{WBGT}_n T_n}{\sum_{i=1}^{n} T_1 + T_2 + \cdots + T_n}$$

$$\text{WBGT}_{\text{TWA}} = \frac{[134.4][72] + [66.6][210] + [99.3][198]}{72 + 210 + 198}$$

$$\text{WBGT}_{\text{TWA}} = \frac{9,676.8 + 13,986.0 + 19,661.4}{480} = \frac{43,324.2}{480} = 90.26$$

∴ The WBGT$_{\text{TWA}}$ for a Typical Open Hearth Operator = 90.3°F = 32.4°C.

Problem #71:

The solution to this problem will use some of the data developed in solving the previous problem (ie. the three WBGT$_{\text{Inside}}$ Indices), entering these values into **Equation #16A** from Page 3-35 to develop the desired solution; thus:

$$\text{WBGT}_{\text{TWA}} = \frac{\sum\limits_{i=1}^{n} \text{WBGT}_i T_i}{\sum\limits_{i=1}^{n} T_i} = \frac{\sum\limits_{i=1}^{n} \text{WBGT}_1 T_1 + \text{WBGT}_2 T_2 + \cdots + \text{WBGT}_n T_n}{\sum\limits_{i=1}^{n} T_1 + T_2 + \cdots + T_n}$$

$$\text{WBGT}_{\text{TWA}} = \frac{[134.4][60] + [66.6][225] + [99.3][195]}{60 + 225 + 195}$$

$$\text{WBGT}_{\text{TWA}} = \frac{8,064.0 + 14,985.0 + 19,363.5}{480} = \frac{42,412.5}{480} = 88.36$$

∴ The Improved WBGT$_{\text{TWA}}$ for a Typical Open Hearth Operator recommended by the Industrial Hygienist = 88.4°F = 31.3°C, an improvement of only 1.9°F = 1.1°C.

Problem #72:

To solve this problem, we must use **Equation #17A** on Page 3-36, thus:

$$V = 49\sqrt{t + 459}$$

$$V_{129°F} = 49\sqrt{129 + 459} = 49\sqrt{588}$$

$$V_{129°F} = [49][24.25] = 1,188.19$$

∴ The Speed of Sound at 129°F in the Mojave Desert is 1,188.2 ft/sec.

Problem #73:

To solve this problem, we must again use **Equation #17A** on Page 3-36, thus:

$$V = 49\sqrt{t + 459}$$

We must, of course, first convert this temperature, given in °C, to its corresponding temperature, in °F. In this case, - 35°C = - 31°F; thus:

$$V_{-35°C} = 49\sqrt{-31 + 459} = 49\sqrt{428}$$

$$V_{-35°C} = [49][20.69] = 1,013.72$$

∴ The Speed of Sound at -35°C in Anchorage, AK, is 1,013.7 ft/sec.

Problem #74:

This problem relies upon the Definition of an Analog Sound Pressure Level, listed as **Equation #18A** on Page 3-36; thus:

$$L_P = 20\,log\,P + 93.98$$

$$115 = 20\,log\,P + 93.98$$

$$115 - 93.98 = 20\,log\,P = 21.02$$

$$log\,P = \frac{21.02}{20} = 1.05$$

& taking the antilog of each side of the equation:

$$P = 11.25$$

∴ The Analog Sound Pressure Level of a 115 dB sound = 11.3 nt/m².

Problem #75:

This problem relies upon the Definition of an Analog Sound Intensity Level, listed as **Equation #18B** on Page 3-37; thus:

$$L_I = 10\,log\,I + 120$$

$$L_I = 10\,log\left[2.45 \times 10^{-7}\right] + 120$$

$$L_I = [10][-6.611] + 120 = -66.11 + 120$$

$$L_I = 53.89$$

> ∴ The Analog Sound Intensity Level of a hovering hummingbird at a distance of 2 meters is 53.9 dB.

Problem #76:

This problem relies upon the Definition of an Analog Sound Power Level, listed as **Equation #18C** on Page 3-37; thus:

$$L_P = 10 \log P + 120$$

$$134 = 10 \log P + 120$$

$$10 \log P = 134 - 120 = 14$$

$$\log P = \frac{14}{10} = 1.4$$

& taking the antilog of each side of the equation, we conclude:

$$P = 25.12$$

> ∴ The Sound Power Level of a Top Fuel Dragster (at maximum engine and super-charger RPM) is 25.1 watts.

Problem #77:

The solution to this problem will require the use of **Equation #19A** from Page 3-38. We must first observe that this aircraft's jet engine is producing sound as a "single hemisphere" radiating source — ie. its jet engine radiates sound <u>only</u> into the air, <u>not</u> into the ground (therefore, <u>only</u> into a single hemisphere); conse-quently, we must use a directionality factor of "**2**"; thus:

$$L_{P-\text{Effective}} = L_{P-\text{Source}} - 20 \log r - 0.5 + 10 \log Q$$

$$L_{P-\text{Effective}} = 165 - 20 \log 300 - 0.5 + 10 \log 2$$

$$L_{P-\text{Effective}} = 165 - [20][2.477] - 0.5 + [10][0.301]$$

$$L_{P-\text{Effective}} = 165 - 49.54 - 0.5 + 3.01 = 117.97$$

∴ The Effective Sound Pressure Level experienced by the Ground Observer listed in this problem would be 118 dB — a very uncomfortably loud sound.

Problem #78:

The solution to this problem will also require the use of **Equation #19A** from Page 3-38. In this case, we must observe that this aircraft's jet engine is producing sound as a "spherical omnidirectional" radiating source — ie. its jet engine is radiating sound in <u>all</u> directions, into the air through which it is moving; consequently, we will use a directionality factor of "**1**"; thus:

$$L_{P-\text{Effective}} = L_{P-\text{Source}} - 20 \log r - 0.5 + 10 \log Q$$

$$117.97 = 165 - 20 \log r - 0.5 + 10 \log 1$$

$$20 \log r = 165 - 117.97 - 0.5 + [10][0] = 46.53$$

$$\log r = \frac{46.53}{20} = 2.33$$

$$r = 212.14$$

∴ The F8U will deliver a ~118 dB sound when it is at an altitude of approximately 212 feet, directly over the Ground Observer.

Problem #79:

This is the classic problem involving the addition of several different quantified (in dB) noise sources with the goal of obtaining a single overall equivalent noise source. It will require the use of **Equation #20A** from Page 3-39, thus:

$$L_{\text{Total}} = 10 \log \left[\sum_{i=1}^{n} 10^{L_i/10} \right] = 10 \log \left[10^{L_1/10} + 10^{L_2/10} + \dots + 10^{L_n/10} \right]$$

$$L_{Total} = 10\,log\left[\left(10^{106/10}\right)(6)\right] = 10\,log\left[(3.981 \times 10^{10})(6)\right]$$

$$L_{Total} = 10\,log\left[2.39 \times 10^{11}\right] = [10][11.38] = 113.78$$

> ∴ The Foreman of this Machine Shop will experience noise, at a combined level of 113.8 dB, from the simultaneous operation of all six grinders.

Problem #80:

This problem is completely analogous to the previous one; it deals with the combined effect of several different sound sources; it, too, will require the use of **Equation #20A** from Page 3-39; thus:

$$L_{Total} = 10\,log\left[\sum_{i=1}^{n} 10^{L_i/10}\right] = 10\,log\left[10^{L_1/10} + 10^{L_2/10} + \cdots + 10^{L_n/10}\right]$$

$$109 = 10\,log\left[(20)\left(10^{L_{Bagpipe}/10}\right)\right]$$

$$log\left[(20)\left(10^{L_{Bagpipe}/10}\right)\right] = \frac{109}{10} = 10.9$$

Exponentiating (to the power of **10**) both sides of this Equation:

$$(20)\left(10^{L_{Bagpipe}/10}\right) = 7.943 \times 10^{10}$$

$$10^{L_{Bagpipe}/10} = \frac{7.943 \times 10^{10}}{20} = 3.972 \times 10^9$$

Taking the common logarithm of both sides of this Equation:

$$\frac{L_{Bagpipe}}{10} = 9.599$$

$$L_{Bagpipe} = [10][9.599] = 95.99$$

∴ Each Bagpipe produces Music/Noise (??) at approximately 96 dB.

Problem #81:

To answer this question, we must first determine the OSHA permitted duration, in hours, for each of the two identified noise levels. Once this has been accomplished, we simply employ **Equation #21A**, from Page 3-40, to obtain the requested result; thus:

$$T_{max} = \frac{8}{2^{L-90/5}}$$

1. For a Noise Level of 104 dB:

$$T_{max\ @\ 104dB} = \frac{8}{2^{[104-90]/5}} = \frac{8}{2^{14/5}}$$

$$T_{max\ @\ 104dB} = \frac{8}{2^{2.8}} = \frac{8}{6.964} = 1.149 \text{ hours}$$

2. For a Noise Level of 92 dB:

$$T_{max\ @\ 92dB} = \frac{8}{2^{[92-90]/5}} = \frac{8}{2^{2/5}}$$

$$T_{max\ @\ 92dB} = \frac{8}{2^{0.4}} = \frac{8}{1.32} = 6.063 \text{ hours}$$

The additional time permitted, ΔT_{max}, (at the lesser Noise Level) is simply the difference between these two OSHA permitted time intervals; thus:

$$\Delta T_{max} = 6.063 - 1.149 = 4.914$$

∴ This Individual can spend an additional 4.9 hours (= 4 hours and 54 minutes) at a 92 dB Noise Level than would be permitted at a 104 dB Noise Level.

Problem #82:

This problem can be solved by using **Equation #21A** from Page 3-40, first to identify the effective Equivalent Sound Pressure Level that is being experienced by the Band Saw Operator when she is wearing her ear plugs, then reapply the same Equation to determine the additional time permitted when she uses ear muffs; thus:

$$T_{max} = \frac{8}{2^{L-90/5}}$$

$$4.6 = \frac{8}{2^{[L-90]/5}} \quad \text{\& transposing this Equation:}$$

$$2^{[L-90]/5} = \frac{8}{4.6} = 1.739 \quad \text{\& taking the common logarithm of both sides}$$

$$log\left[2^{[L-90]/5} \right] = log\,1.739$$

$$[0.301]\left[\frac{L-90}{5} \right] = 0.240$$

$$\frac{L-90}{5} = \frac{0.240}{0.301} = 0.798$$

$$L-90 = [0.798][5] = 3.992$$

$$L = 3.992 + 90 = 93.992$$

Therefore, this Band Saw Operator experiences a Noise Level of 94 dB while she operates the band saw wearing ear plugs. If she wears ear muffs, she will experience a Noise Level 7 dB lower than 94 dB (ie. there is a 31 dB reduction with ear muffs vs. 24 dB reduction with ear plugs). The new reduced Noise Level will, therefore, become 87 dB, and the time permitted at this level will be given by **Equation #21A**, as follows:

$$T_{max} = \frac{8}{2^{L-90/5}}$$

$$T_{max\,@\,87dB} = \frac{8}{2^{[87-90]/5}} = \frac{8}{2^{-3/5}}$$

$$T_{max \ @ \ 87dB} = \frac{8}{2^{-0.6}} = \frac{8}{0.660} = 12.126$$

∴ Using ear muffs, this Band Saw Operator will be able to operate her band saw, without the danger of suffering any hearing loss, for up to 12.1 hours (= 12 hours and 6 minutes) per day.

Problem #83:

The solution to this problem will require the use of both **Equation #21A** from Page 3-40, and **Equation #21B** from Page 3-41, and in that order; thus:

$$T_{max} = \frac{8}{2^{L-90/5}}$$

1. For an average 95 dB SPL:

$$T_{max \ @ \ 95dB} = \frac{8}{2^{[95-90]/5}} = \frac{8}{2^{5/5}}$$

$$T_{max \ @ \ 95dB} = \frac{8}{2^{1.0}} = \frac{8}{2.00} = 4.00 \ hours$$

2. For an average 90 dB SPL:

$$T_{max \ @ \ 90dB} = \frac{8}{2^{[90-90]/5}} = \frac{8}{2^{0/5}}$$

$$T_{max \ @ \ 90dB} = \frac{8}{2^{0.0}} = \frac{8}{1.00} = 8.00 \ hours$$

3. For an average 84 dB SPL:

$$T_{max \ @ \ 84dB} = \frac{8}{2^{[84-90]/5}} = \frac{8}{2^{-6/5}}$$

$$T_{max \ @ \ 84dB} = \frac{8}{2^{-1.2}} = \frac{8}{0.435} = 18.379 \ hours$$

4. For an average 82 dB SPL:
$$T_{max \, @ \, 82dB} = \frac{8}{2^{[82-90]/5}} = \frac{8}{2^{-8/5}}$$

$$T_{max \, @ \, 82dB} = \frac{8}{2^{-1.6}} = \frac{8}{0.330} = 24.251 \text{ hours}$$

Now with these Maximum Time Periods at the various average SPLs, we can apply **Equation #21B** from Page 3-41, to obtain the requested result; thus:

$$D = \sum_{i=1}^{n} \frac{C_i}{T_{max_i}} = \frac{C_1}{T_{max_1}} + \frac{C_2}{T_{max_2}} + \cdots + \frac{C_n}{T_{max_n}}$$

$$D_{Lathe \, Operator} = \frac{1.5}{4.00} + \frac{4.5}{8.00} + \frac{1}{18.379} + \frac{1}{24.251}$$

$$D_{Lathe \, Operator} = 0.375 + .563 + 0.054 + 0.041 = 1.033$$

∴ This Worker's Daily Dose (of Noise), expressed as a percentage, is 103.3%.

Problem #84:

The solution to this problem, which is an extension of the previous one, requires the use of **Equation #21B**, from Page 3-42; thus:

$$L_{equivalent} = 90 + 16.61 \, log \, D$$

$$L_{equivalent} = 90 + 16.61 \, log \, 1.0331$$

$$L_{equivalent} = 90 + [16.61][0.014] = 90 + 0.235$$

$$L_{equivalent} = 90.235$$

∴ This Worker experienced an Equivalent Sound Pressure Level of 90.24 dB.

Problem #85:

Like the Problem #83, earlier, the solution to this problem will require the use of both **Equation #21A** from Page 3-40, and **Equation #21B** from Page 3-41, and in that order; thus:

$$T_{max} = \frac{8}{2^{[L-90]/5}}$$

1. For an average 81 dB SPL:

$$T_{max\ @\ 81dB} = \frac{8}{2^{[81-90]/5}} = \frac{8}{2^{-9/5}}$$

$$T_{max\ @\ 81dB} = \frac{8}{2^{-1.8}} = \frac{8}{0.287} = 27.858 \text{ hours}$$

2. For an average 93 dB SPL:

$$T_{max\ @\ 93dB} = \frac{8}{2^{[93-90]/5}} = \frac{8}{2^{3/5}}$$

$$T_{max\ @\ 93dB} = \frac{8}{2^{0.6}} = \frac{8}{1.516} = 5.278 \text{ hours}$$

3. For an average 96 dB SPL:

$$T_{max\ @\ 96dB} = \frac{8}{2^{[96-90]/5}} = \frac{8}{2^{6/5}}$$

$$T_{max\ @\ 96dB} = \frac{8}{2^{1.2}} = \frac{8}{2.297} = 3.482 \text{ hours}$$

4. For an average 98 dB SPL:

$$T_{max\ @\ 98dB} = \frac{8}{2^{[98-90]/5}} = \frac{8}{2^{8/5}}$$

$$T_{max\ @\ 98dB} = \frac{8}{2^{1.6}} = \frac{8}{3.031} = 2.639 \text{ hours}$$

Now with these Maximum Time Periods at the various average SPLs, we can apply **Equation #21B** from Page 3-41, to obtain the requested result; thus:

$$D = \sum_{i=1}^{n} \frac{C_i}{T_{max_i}} = \frac{C_1}{T_{max_1}} + \frac{C_2}{T_{max_2}} + \cdots + \frac{C_n}{T_{max_n}}$$

$$D_{Printer} = \frac{4.5}{27.858} + \frac{2.1}{5.278} + \frac{1.0}{3.482} + \frac{0.4}{2.639}$$

$$D_{Printer} = 0.162 + 0.398 + 0.287 + 0.152 = 0.998$$

∴ These Printers' Daily Dose (of Noise), expressed as a percentage = 99.8%.

& The Printing Company that employs these four Printers is not in violation of any established OSHA Sound Pressure Level Dosage Standards.

Problem #86:

The solution to this problem, which is an extension of the previous one, will require the use of **Equation #21B**, from Page 3-42; thus:

$$L_{equivalent} = 90 + 16.61 \log D$$

$$L_{equivalent} = 90 + 16.61 \log 0.998$$

$$L_{equivalent} = 90 + [16.61][-0.001] = 90 - 0.013 = 89.987$$

∴ These Printers experience an Equivalent Sound Pressure Level of 89.99 dB.

Problem #87:

The solution to this problem requires only the Descriptive Definition of the A-Scale Weighting Factors, as shown in the Tabulation associated with **Definition #22A** from Page 3-42. Clearly this note ("C-below-Middle-C", reference the piano scale) falls into the 250 Hz Octave Band. For this Octave Band, we must deduct 9 dB from every identified Linear Sound Pressure level; thus:

∴ The SPL of this "C-Below-Middle-C" Tuning Fork = 71 dB$_{Linear}$ = 62 dBA.

Problem #88:

Again from the Descriptive Definition of the A-Scale Weighting Factors, as shown in the Tabulation associated with **Definition #22A** from Page 3-42, we can see that the only Octave Band that does not have a SPL adjustment is the 1,000 Hz Octave Band. To determine the Upper and Lower Band Edge Frequencies, we must employ the two Equations associated with **Definition #22B** from Page 3-43; thus:

$$f_{upper-1/1} = 2f_{lower-1/1}$$

$$f_{center-1/1} = \sqrt{f_{upper-1/1} f_{lower-1/1}} = \sqrt{2}\left[f_{lower-1/1}\right]$$

$$1,000 = \sqrt{2}\left[f_{lower-1/1}\right]$$

$$f_{lower-1/1} = \frac{1,000}{\sqrt{2}} = \frac{1,000}{1.414} = 707.1$$

$$f_{upper-1/1} = [2][707.1] = 1,414.2$$

∴ The Upper and Lower Band Edge Frequencies of the 1,000 Hz Standard Unitary Octave Band are as follows:
Upper Band Edge Frequency = 1,414 Hz
Lower Band Edge Frequency = 707 Hz

Problem #89:

To determine the Center Frequency of the Octave Band for which the Lower Band Edge Frequency is known, we must employ the two Equations associated with **Definition #22B**, taken from Page 3-43; thus:

$$f_{upper-1/1} = 2f_{lower-1/1}$$

$$f_{center-1/1} = \sqrt{f_{upper-1/1} f_{lower-1/1}} = \sqrt{2}\left[f_{lower-1/1}\right]$$

We must first determine the Upper Band Edge Frequency, and then use this information to determine the Center Frequency; thus:

$$f_{upper-1/1} = [2][2,828] = 5,656$$

$$f_{center-1/1} = \sqrt{[2,828][5,656]} = \sqrt{15,995,168} = 3,999.4$$

∴ The Center Frequency is 4,000 Hz (rounding up from the result of 3,999.4 Hz), and this is the Standard 4,000 Hz = 4 kHz Octave Band.

Problem #90:

To solve this problem, we will have to apply the two Equations that make up the relationships for Half Octave Bands, as shown in **Equation #22C**, on Page 3-44; thus:

$$f_{upper-1/2} = \sqrt{2}f_{lower-1/2}$$

$$f_{center-1/2} = \sqrt{f_{upper-1/2}f_{lower-1/2}}$$

$$f_{center-1/2} = \sqrt{\sqrt{2}f_{lower-1/2}f_{lower-1/2}} = f_{lower-1/2}\sqrt{\sqrt{2}}$$

$$354 = f_{lower-1/2}\sqrt[4]{2} = [1.189]f_{lower-1/2}$$

$$f_{lower-1/2} = \frac{354}{1.189} = 297.7$$

$$f_{upper-1/2} = \sqrt{2}[297.7] = [1.414][297.7] = 421.0$$

∴ The Upper and Lower Band Edge Frequencies of the 354 Hz Standard Half Octave Band are as follows:
    Upper Band Edge Frequency = ~ 421 Hz
    Lower Band Edge Frequency = ~ 298 Hz

Problem #91:

To solve this problem, we will have to apply the two Equations that make up the relationships for any 1/nth Octave Bands, as shown in **Equation #22D**, on Page 3-45. We must start out by determining the Upper Band Edge Frequency for this 1/3rd Octave Band; thus:

$$f_{upper-1/n} = \sqrt[n]{2}\, f_{lower-1/n}$$

$$f_{center-1/n} = \sqrt{f_{upper-1/n}\, f_{lower-1/n}}$$

$$f_{upper-1/3} = \sqrt[3]{2}\, f_{lower-1/3}$$

$$f_{upper-1/3} = \sqrt[3]{2}\,[1,122] = [1.260][1,122] = 1,413.6$$

$$f_{center-1/3} = \sqrt{[1,122][1,413.6]} = \sqrt{1,586,094.5} = 1,259.4$$

∴ The Standard 1/3rd Octave Band, for which the Lower Band Edge Frequency is 1,122 Hz, has a Center Frequency of 1,259 Hz.

Problem #92:

To solve this problem, we must use what is probably the most basic relationship in all of ventilation engineering, namely, **Equation #23A** from Page 3-46; thus:

$$Q = AV$$

We must first transpose this relationship into a format that better suits the data provided in this problem, namely, into a format that will accommodate a duct diameter measured in "inches" and a velocity measured in "fpm"; thus:

$$Q = \frac{\pi D^2 V}{4} = \frac{\pi d^2 V}{576}$$

$$Q = \frac{\pi[10]^2[2,500]}{576} = \frac{\pi[100][2,500]}{576}$$

$$Q = \frac{785,398.16}{576} = 1,363.5$$

∴ The Volumetric Flow Rate in this Duct is 1,364 cfm.

Problem #93:

This problem uses the same relationship as did the former problem, namely, **Equation #23A** from Page 3-46; thus:

$$Q = AV \quad \& \quad Q = \frac{\pi d^2 V}{576}$$

Transposing to solve for "d", we get:

$$d = \sqrt{\frac{576Q}{\pi V}}$$

$$d = \sqrt{\frac{[576][25,135]}{\pi[8,000]}} = \sqrt{\frac{14,477,760}{25,132.74}}$$

$$d = \sqrt{576.05} = 24.0$$

∴ The Diameter of this HVAC Duct is 24 inches.

Problem #94:

This extension to **Problem #93** requires the use of **Equation #23B** on Page 3-46 to develop its solution; thus:

$$V = 1,096\sqrt{\frac{VP}{\rho}}$$

Transposing to solve for "VP", we get:

$$VP = \frac{V^2 \rho}{[1,096]^2}$$

$$VP = \frac{[8,000]^2 [0.075]}{[1,096]^2}$$

$$VP = \frac{4,800,000}{1,201,216} = 4.00$$

∴ The Velocity Pressure in this 24 inch diameter HVAC Duct is 4.0 inches of water.

Problem #95:

To solve this problem, we must use **Equation #23C** from Page 3-47; thus:

$$V = 4,005\sqrt{VP}$$

$$V = 4,005\sqrt{2.32} = [4,005][1.523] = 6,100.2$$

∴ The Duct Velocity in this Duct is 6,100 fpm.

Problem #96:

This problem, too, requires the use of **Equation #23C** from Page 3-47; thus:

$$V = 4,005\sqrt{VP}$$

$$V = 4,005\sqrt{0.45} = [4,005][0.671] = 2,686.6$$

∴ The Duct Velocity in this 8 inch Diameter Duct is 2,687 fpm.

Problem #97:

This problem also requires the use of **Equation #23C** from Page 3-47. We must first convert the listed Duct Velocity, which has been given in ft/sec, to the required format in fpm; thus:

$$V = 4,005 \sqrt{VP}$$

Transposing to solve for "VP", we get:

$$VP = \left[\frac{V}{4,005} \right]^2$$

$$VP = \left[\frac{[25][60]}{4,005} \right]^2 = \left[\frac{1,500}{4,005} \right]^2 = [0.375]^2 = 0.140$$

∴ The calculated Velocity Pressure under the stipulated conditions is 0.14 inches of water. I would guess that the recent graduate could stand to improve on his skill in the empirical determination of Velocity Pressures in ducts.

Problem #98:

This problem requires the application of **Equation #23E** from Page 3-48. Since we are dealing with a duct at its exhaust point, we can safely assume that the actual Static Pressure must be positive (remember that Static Pressure can be either positive or negative). If this Static Pressure were to have been negative, air would have been flowing into the duct at this point, rather than exhausting as we have been told; therefore, we must assume that the stated Static Pressure is +0.01 inches of water; thus:

$$TP = SP + VP$$

Transposing to solve for "VP", we get:

$$VP = TP - SP$$

$$VP = 0.41 - 0.01 = 0.40$$

∴ The Velocity Pressure at this Duct Exhaust Point is + 0.40 inches of water.

Problem #99:

The solution to this problem, too, will require the use of **Equation #23E** from Page 3-48. In this case, we see that the Total Pressure has been reported to be 0.24 inches of water; however, we must note that the point at which this measurement has been made is at the inlet to a fan. Because of this, we must assume that this Total Pressure is negative, ie. TP = - 0.24 inches of water, since that is the characteristic of this measurement at the inlet to <u>any</u> fan. Remember that Velocity Pressure is always positive; thus:

$$TP = SP + VP$$

Transposing to solve for "SP", we get:

$$SP = TP - VP$$

$$SP = -0.24 - 0.55 = -0.79$$

∴ The Static Pressure at the Inlet to this Fan is - 0.79 inches of water.

Problem #100:

This extension to the foregoing problem will require the use of **Equation #23F** on Page 3-48; thus:

$$SP_1 + VP_1 = SP_2 + VP_2 + \text{losses}$$

In order to apply this relationship, we will first have to determine the "losses". We are given that these losses are frictional in nature and occur at a rate of 0.02 inches of water/20 lineal feet of duct. We have also been told that there are 220 feet of duct between the fan inlet and the inlet end of the duct at a hood. We can therefore determine the "losses" as follows:

$$\text{losses} = \left\lfloor \frac{220}{20} \right\rfloor [0.02] = [11][0.023] = 0.22$$

For the purposes of **Equation #23F** that we will use to develop the requested answer to this problem, we can assume that the Velocity Pressures at both points in the duct (ie. at the duct inlet in the hood, and at the fan inlet) are equal, since the gas if flowing at the same velocity at all points in this single duct; thus:

$$SP_1 = -0.79 + 0.22 = -0.57$$

> ∴ The Static Pressure at the Hood Inlet would be - 0.57 inches of water.

Problem #101:

The solution to this problem also requires the use of **Equation #23F** on Page 3-48; thus:

$$SP_1 + VP_1 = SP_2 + VP_2 + \text{losses}$$

Again, we can assume that the Velocity Pressures are the same at each of the two points where the Air Balancing Technician made measurements of Static Pressure; thus:

$$- 0.35 = - 0.41 + \text{losses}$$

$$\text{losses} = - 0.35 + 0.41 = 0.06$$

Therefore, the total losses in this 75 foot section of duct are 0.06 inches of water; and the specific frictional losses per 10 lineal feet, expressed as L, are given by:

$$\text{losses} = \left|\frac{\text{length of duct}}{\text{specific lineal measure}}\right| L$$

Transposing to solve for "L", we get:

$$L = \left\lfloor \frac{10}{75} \right\rfloor [0.06] = 0.008$$

> ∴ The Frictional Losses in this Duct are at a rate of 0.008 inches of water/10 lineal feet of Duct.

Problem #102:

The solution to this problem depends upon **Equation #24A** from Page 3-49 (note, the Hood Opening in this situation is <u>Unflanged</u>); thus:

$$V = \frac{Q}{10x^2 + A}$$

Transposing to solve for "Q", we get:

$$Q = V\left[10x^2 + A\right]$$

$$Q = [100]\left[(10)(2.5)^2 + (1)(3)\right]$$

$$Q = [100][62.5 + 3] = [100][65.5] = 6,550$$

∴ The Minimum Volumetric Flow Rate required to affect a Capture of the Irritating Vapor is 6,550 cfm.

Problem #103:

This variation on the previous problem will employ **Equation #24B**, also on Page 3-49. In this case, we are dealing with a Flanged Hood Opening; thus:

$$V = \frac{Q}{0.75\left[10x^2 + A\right]}$$

Transposing to solve for "x", we get — after several steps:

$$10x^2 + A = \frac{Q}{0.75V}$$

$$10x^2 = \frac{Q}{0.75V} - A = \frac{Q - 0.75VA}{0.75V}$$

$$x^2 = \frac{Q - 0.75VA}{[10][0.75V]} = \frac{Q - 0.75VA}{7.5V} \quad \text{\& finally:}$$

$$x = \sqrt{\frac{Q - 0.75VA}{7.5V}}$$

$$x = \sqrt{\frac{6,550 - [0.75][100][3][1]}{[7.5][100]}}$$

$$x = \sqrt{\frac{6,550 - 225}{750}} = \sqrt{\frac{6,325}{750}}$$

$$x = \sqrt{8.433} = 2.904$$

The distance, "d", that the irritating vapor source could be moved back away from the newly Flanged Hood opening, is given by the difference between the just calculated source distance, 2.90 feet, and the original distance of 2.50 feet, as listed in **Problem #102**, in the situation wherein the Hood had not been equipped with a flanged opening, and is given by:

$$d = 2.90 - 2.50 = 0.40$$

∴ The Vapor Source could be moved back — away from its current location in front of the now Flanged Hood Opening — by a total of 0.40 feet ~ 4.9 inches.

Problem #104:

To solve this problem, we must again use **Equation #24B** from Page 3-49; thus:

$$V = \frac{Q}{0.75\left[10x^2 + A\right]}$$

It must be remembered that the Area Term in the denominator of this Equation, namely, "A", must be in units of "square feet", thus we must make the conversion from the units provided in the Problem Statement.

$$V = \frac{8,500}{0.75\left[10(3)^2 + (0.25)(1)\right]}$$

$$V = \frac{8,500}{0.75[90 + 0.25]} = \frac{8,500}{[0.75][90.25]}$$

$$V = \frac{8,500}{67.688} = 125.58$$

We have been told that a Capture Velocity of 2 fps is required to be able to acquire the respirable particulates. Expressing this in feet per minute, we see that the required Capture Velocity is 120 fpm. Clearly 125.58 fpm is greater than 120 fpm.

∴ The Centerline Velocity of this Flanged Hood System will be 125.6 fpm, when it is operated in the manner described. It appears that this Hood System will, indeed, operate satisfactorily in the current situation.

Problem #105:

To solve this problem, we will have to employ **Equation #24C** from Page 3-50; thus:

$$SP_h = VP_d + h_e$$

$$SP_h = 15.2 + 8.1 = 23.3$$

∴ The Hood Static Pressure for this situation is 23.3 inches of water.

Problem #106:

This extension of **Problem #105** will require the use of **Equation #24D**, also from Page 3-50; thus:

$$C_e = \sqrt{\frac{VP_d}{SP_h}}$$

$$C_e = \sqrt{\frac{15.2}{23.3}} = \sqrt{0.652} = 0.808$$

∴ The Coefficient of Entry for this Hood is 0.81.

Problem #107:

To solve this problem, we must first use **Equation #23E** on Page 3-48, to obtain the Velocity Pressure in the Duct, and then use this piece of information to obtain the required answer, using **Equation #24C** on Page 3-50; thus:

$$TP = SP + VP_d$$

$$6.2 = -1.2 + VP_d$$

$$VP_d = 6.2 + 1.2 = 7.4$$

Now we can apply **Equation #24C** from Page 3-50 to obtain the requested result:

$$SP_h = VP_d + h_e$$

$$12.5 = 7.4 + h_e$$

$$h_e = 12.5 - 7.4 = 5.1$$

∴ The Hood Entry Losses for this Hood are 5.1 inches of water.

Problem #108:

The solution to this extension of **Problem #107** requires the use of **Equation #24D** from Page 3-50; thus:

$$C_e = \sqrt{\frac{VP_d}{SP_h}}$$

$$C_e = \sqrt{\frac{7.4}{12.5}} = \sqrt{0.592} = 0.769$$

∴ The Coefficient of Entry for this Hood is 0.77.

Problem #109:

The solution to this problem will require the use of **Equation #24E**, the Hood Throat Suction Equation, from Page 3-51; thus:

$$Q = 4,005\,AC_e\,\sqrt{SP_h}$$

Before proceeding, we must make several observations. First, the duct Static Pressure measurement of - 1.75 inches of water was made 12 inches deep into this duct, as measured from its unflanged opening. Since we are dealing with a 4 inch diameter duct, we can see that this Static Pressure measurement was made <u>three</u> full duct diameters downstream from the previously mentioned duct opening. We can safely assume, therefore, that this measurement was downstream of any possible vena contracta that might exist in the duct; thus we can consider that the duct Static Pressure determined at this point will be equal to the Hood Static Pressure, SP_h. We can, therefore, proceed with substitution into **Equation #24E**; thus:

$$Q = 4,005\,AC_e\,\sqrt{SP_h}$$

Note, the Area Term, A, in this Equation must be in "square feet", thus we must convert the diameter that has been provided in inches into its equivalent in "feet".

$$Q = [4,005][\pi]\left[\frac{1}{6}\right]^2 [0.72]\sqrt{1.75}$$

$$Q = \frac{[4,005][\pi][0.72]\sqrt{1.75}}{[6]^2} = \frac{[4,005][\pi][0.72][1.323]}{36}$$

$$Q = \frac{11,984.1}{36} = 332.9$$

∴ The Volumetric Flow Rate for this Hood is approximately 333 cfm.

Problem #110:

To solve this problem, which is an extension of **Problem #108**, we must apply **Equation #23A** from Page 3-46; thus:

$$Q = AV$$

Modifying this Equation to produce one that will employ the information given in the Problem statement directly, we get:

$$V = \frac{Q}{A} = \frac{4Q}{\pi D^2}$$

If we consider that the diameter is measured in inches — rather than feet — and if we let "d" be the value of this diameter in inches, we get:

$$V = \frac{576Q}{\pi d^2}$$

$$V = \frac{[576][332.9]}{\pi[4]^2} = \frac{[576][332.9]}{\pi[16]}$$

$$V = \frac{191,744.9}{50.265} = 3,814.64$$

∴ The Duct Velocity for the Hood System described in this Problem = 3,815 fpm.

Problem #111:

To solve this problem and obtain the average Duct Velocity Pressure, we must go back to **Equation #23C** from Page 3-47; thus:

$$V = 4,005\sqrt{VP}$$

Transposing to solve for "VP", we get:

$$VP = \left[\frac{V}{4,005}\right]^2$$

$$VP = \left[\frac{3,814.6}{4,005}\right]^2 = [0.952]^2 = 0.907$$

∴ The Average Duct Velocity Pressure for this Hood is 0.91 inches of water.

Problem #112:

The solution to this problem will rely, initially, upon **Equation #24E** from Page 3-51, to obtain the Duct Velocity Pressure, then with this information we will use **Equation #23E** from Page 3-48 to develop the requested result; thus:

$$Q = 4,005\,A\sqrt{VP_d}$$

Transposing to solve for "VP_d", we get:

$$VP_d = \left[\frac{Q}{4,005\,A}\right]^2$$

Again, note that the Area Term, "A", is in "square feet", thus we must convert the dimensions of the hood opening that were provided in inches in the Problem Statement into their equivalent length in feet, thus:

$$VP_d = \left[\frac{3,000}{(4,005)(0.5)^2}\right]^2 = \left[\frac{3,000}{(4,005)(0.25)}\right]^2$$

$$VP_d = \left[\frac{3,000}{1,001.25}\right]^2 = [2.996]^2 = 8.978 \sim 9.0$$

Now, with this information, we can determine the requested Duct Total Pressure, using **Equation #23E** from Page 3-48, as stated above; thus:

$$TP = SP + VP$$

$$TP = -4.5 + 9.0 = 4.5$$

∴ The Duct Total Pressure for this Hood is 4.5 inches of water.

Problem #113:

To solve this problem, we must employ the Hood Entry Loss Equation, **Equation #24F** from Page 3-52; thus:

$$h_e = \left|\frac{1 - C_e^2}{C_e^2}\right|VP_h$$

$$h_e = \left|\frac{1 - (0.85)^2}{(0.85)^2}\right|[21.0] = \left[\frac{1 - 0.723}{0.723}\right][21.0]$$

$$h_e = \left\lfloor \frac{0.278}{0.723} \right\rfloor [21.0] = [0.384][21.0] = 8.066$$

∴ The Hood Entry Losses for this Hood System are 8.1 inches of water.

Problem #114:

This problem, too, requires the use of **Equation #24F** from Page 3-52; however, in this case we shall have to begin by transposing this relationship so as to solve for the Coefficient of Entry term; thus:

$$h_e = \left\lfloor \frac{1 - C_e^2}{C_e^2} \right\rfloor VP_h \quad \text{now transposing & solving for "}C_e\text{", as stated above:}$$

$$C_e^2 h_e = \left[1 - C_e^2 \right] VP_h = VP_h - C_e^2 VP_h$$

$$C_e^2 h_e + C_e^2 VP_h = VP_h = C_e^2 \left[h_e + VP_h \right]$$

$$C_e^2 = \frac{VP_h}{h_e + VP_h}$$

$$C_e = \sqrt{\frac{VP_h}{h_e + VP_h}}$$

$$C_e = \sqrt{\frac{12.3}{6.5 + 12.3}} = \sqrt{\frac{12.3}{18.8}} = \sqrt{0.654} = 0.809$$

∴ The Coefficient of Entry for this particular Hood = 0.81.

Problem #115:

The solution to this problem relies upon the Hood Entry Loss Factor Equation, **Equation #24G** from Page 3-52; thus:

$$F_h = \frac{h_e}{VP_h}$$

$$F_h = \frac{6.5}{12.3} = 0.528$$

∴ The Hood Entry Loss Factor for this particular Hood = 0.53.

Problem #116:

The solution to this problem will rely upon the Compound Hood Equations, which are listed on Page 3-53. In this case, we have been given that the Duct Velocity is greater than the Slot Velocity; therefore, we can conclude that the principal relationship we will be using to develop the solution to this Problem will be **Equation #24H**. We must initially start out be determining the Duct Velocity Pressure, and we can do this by employing **Equation #23C** from Page 3-47; thus:

$$V_d = 4,005\sqrt{VP_d}$$

Transposing and solving for "VP_d", we get:

$$VP_d = \left[\frac{V_d}{4,005}\right]^2$$

$$VP_d = \left[\frac{5,600}{4,005}\right]^2 = [1.398]^2 = 1.955 \sim 1.96$$

We must next employ the same relationship to determine the Slot Velocity Pressure; thus:

$$VP_s = \left[\frac{V_s}{4,005}\right]^2$$

$$VP_s = \left[\frac{3,400}{4,005}\right]^2 = [0.849]^2 = 0.721 \sim 0.72$$

Now using these two Velocity Pressure results, coupled with the two approximations that apply to (1) the Slot and (2) the Duct Entry Losses, as listed under **Equation #'s 24H & 24I**, on Page 3-53, we have:

$$h_{ES} = 1.78[VP_s] \quad \& \quad h_{ED} = 0.25[VP_d]$$

$$h_{ES} = [1.78][0.721] = 1.283$$

$$h_{ED} = [0.25][1.955] = 0.489$$

Now, finally, we can employ **Equation #24H** from Page 3-53, to determine the Hood Static Pressure; thus:

$$SP_h = h_{ES} + h_{ED} + VP_d \quad \text{for} \quad V_d > V_s$$

$$SP_h = 1.283 + 0.489 + 1.955 = 3.727$$

∴ For this Compound Hood, the Hood Static Pressure = 3.7 inches of water.

Problem #117:

This is a slightly more complex problem that will, like its predecessor, employ one of the Compound Hood Equations to determine the Hood Static Pressure. In order to determine which of these two relationships we employ, we must first calculate some of the other parameters so as to discover whether the Slot Velocity is greater than the Duct Velocity or vice versa. To start on this track, let us use **Equation #23E** from Page 3-48 to determine the Duct Velocity Pressure; thus:

$$TP_d = SP_d + VP_d$$

Transposing to solve for "VP_d", we get:

$$VP_d = TP_d - SP_d$$

$$VP_d = 1.3 - [-1.9] = 1.3 + 1.9 = 3.2$$

Since we now know the Velocity Pressure in this exhaust duct, we can easily determine the Duct Velocity by using **Equation #23C** from Page 3-47; thus:

$$V_d = 4,005\sqrt{VP_d}$$

$$V_d = 4,005\sqrt{3.2} = [4,005][1.789] = 7,164.4$$

Clearly, now, we can see that the Slot Velocity (given as 7,500 fpm) is greater than the Duct Velocity (calculated out to be 7,164 fpm); therefore, we will eventually be using **Equation #24H** from Page 3-53 to determine the Hood Slot Velocity. As with Problem #116, we must begin by determining all of the appropriate Velocity Pressures. We have already calculated the Velocity Pressure in the Duct, so we must now calculate the Velocity Pressure in the Slot, using the transposed form of **Equation #23C** from Page 3-47, which form was developed for calculating this parameter as it was required **Problem #116**; thus:

$$VP_s = \left[\frac{V_s}{4,005}\right]^2$$

$$VP_s = \left[\frac{7,500}{4,005}\right]^2 = [1.873]^2 = 3.507 \sim 3.5$$

Now using these two Velocity Pressure results, coupled with the two approximations that apply to (1) the Slot and (2) the Duct Entry Losses, as listed under **Equation #'s 24H & 24I**, on Page 3-53, we have:

$$h_{ES} = 1.78[VP_s] \quad \& \quad h_{ED} = 0.25[VP_d]$$

$$h_{ES} = [1.78][3.507] = 6.242$$

$$h_{ED} = [0.25][3.2] = 0.80$$

Now, finally, we can employ **Equation #24I** from Page 3-53, to determine the Hood Static Pressure; thus:

$$SP_h = h_{ES} + h_{ED} + VP_s \quad \text{for} \quad V_s > V_d$$

$$SP_h = 6.242 + 0.800 + 3.507 = 10.549$$

∴ The Hood Static Pressure for this Compound Hood = 10.5 inches of water.

Problem #118:

To solve this problem, we must employ the first of the relationships involving the rotational speed of a fan, namely, **Equation #25A** on Page 3-54; thus:

$$\frac{CFM_1}{CFM_2} = \frac{RPM_1}{RPM_2}$$

$$\frac{CFM_1}{10,500} = \frac{1,450}{1,750}$$

$$CFM_1 = \frac{[1,450][10,500]}{1,750} = \frac{15,225,000}{1,750} = 8,700$$

∴ The previous Discharge Volume of this Fan was 8,700 cfm.

Problem #119:

The solution to this problem relies on the second of the Fan Rotational Speed relationships, namely, **Equation #25B** on Page 3-54; thus:

$$\frac{SP_{FAN_1}}{SP_{FAN_2}} = \left[\frac{RPM_1}{RPM_2}\right]^2$$

$$\frac{SP_{FAN_2}}{4.6} = \left[\frac{1,750}{1,450}\right]^2 = [1.207]^2 = 1.457$$

$$SP_{FAN_2} = \left[\frac{1,750}{1,450}\right]^2 [4.6] = [1.457][4.6] = 6.70$$

∴ The new Fan Static Discharge Pressure = 6.7 inches of water.

Problem #120:

The solution to this problem relies on the third of the three relationships involving the Rotational Speeds of Fans, namely, **Equation #25C** on Page 3-55, thus:

$$\frac{BHP_1}{BHP_2} = \left[\frac{RPM_1}{RPM_2}\right]^3$$

$$\frac{BHP_2}{15} = \left[\frac{1,750}{1,450}\right]^3$$

$$BHP_2 = \left[\frac{1,750}{1,450}\right]^3 [15] = [1.207]^3[15]$$

$$BHP_2 = [1.785][15] = 26.37$$

∴ The new fan motor will have a Brake Horsepower rating of 26.4 bhp, minimum.

Problem #121:

The solution to this problem will require the use of **Equation #25A** from Page 3-54; thus:

$$\frac{CFM_1}{CFM_2} = \frac{RPM_1}{RPM_2}$$

$$\frac{RPM_2}{1,200} = \frac{[25,000][1.40]}{25,000} = 1.40$$

$$RPM_2 = [1.40][1,200] = 1,680$$

∴ Under the new Conditions, the Fan Rotational Speed must be 1,680 rpm.

Problem #122:

The solution to this problem will rely on **Equation #25B** on Page 3-54; thus:

$$\frac{SP_{FAN_1}}{SP_{FAN_2}} = \left[\frac{RPM_1}{RPM_2}\right]^2$$

$$\frac{SP_{FAN_2}}{6.25} = \left[\frac{1,680}{1,200}\right]^2$$

$$SP_{FAN_2} = \left[\frac{1.680}{1.200}\right]^2 [6.25] = [1.40]^2 [6.25]$$

$$SP_{FAN_2} = [1.96][6.25] = 12.25$$

∴ The new Fan Static Pressure will be 12.3 inches of water.

Problem #123:

This problem can be solved by using **Equation #25C** from Page 3-55; thus:

$$\frac{BHP_1}{BHP_2} = \left[\frac{RPM_1}{RPM_2}\right]^3$$

$$\frac{BHP_2}{50} = \left[\frac{1,680}{1,200}\right]^3$$

$$BHP_2 = \left[\frac{1.680}{1.200}\right]^3 [50] = [1.40]^3 [50]$$

$$BHP_2 = [2.744][50] = 137.2$$

∴ The Requirements for the Modified Ventilation System call for a new Fan Motor that has a Capacity of at least 137.2 bhp.

Problem #124:

The solution to this problem will rely on the first of the three relationships involving the Diameters of Fans, namely, **Equation #26A** on Page 3-56; thus:

$$\frac{CFM_1}{CFM_2} = \left[\frac{D_1}{D_2}\right]^3$$

Transposing and solving for "D_2", we get:

$$D_2 = D_1 \sqrt[3]{\frac{CFM_2}{CFM_1}}$$

$$D_2 = [6.0] \sqrt[3]{\frac{67,500}{28,500}} = [6.0] \sqrt[3]{2.368}$$

$$D_2 = [6.0][1.333] = 7.998 \sim 8.00$$

$\therefore$ The Industrial Hygienist will likely recommend a completely new Ventilation System employing 8.00 inch diameter components.

Problem #125:

The solution to this problem will rely on **Equation #26B** on Page 3-56; thus:

$$\frac{SP_{FAN_1}}{SP_{FAN_2}} = \left[\frac{D_1}{D_2}\right]^2$$

Transposing and solving for "SP_{FAN_2}", we get:

$$SP_{FAN_2} = SP_{FAN_1} \left[\frac{D_2}{D_1}\right]^2$$

$$SP_{FAN_2} = [7.0] \left[\frac{8.0}{6.0}\right]^2 = [7.0][1.333]^2$$

$$SP_{FAN_2} = [7.0][1.778] = 12.444$$

∴ The New Fan Static Discharge Pressure = 12.4 inches of water.

Problem #126:

This problem will require the application of **Equation #26C** from Page 3-57; thus:

$$\frac{BHP_1}{BHP_2} = \left[\frac{D_1}{D_2}\right]^5$$

Transposing and solving for "BHP$_1$", we get:

$$BHP_1 = BHP_2\left[\frac{D_1}{D_2}\right]^5$$

$$BHP_1 = [150]\left[\frac{6.0}{8.0}\right]^5 = [150][0.75]^5$$

$$BHP_1 = [150][0.237] = 35.60$$

∴ The Brake Horsepower of the previous Fan Motor was approximately 35.6 bhp.

Problem #127:

To solve this problem we must first calculate what the Volumetric Flow Rate was in the 15.0 inch diameter Duct. To do this, we will use **Equation #23A** from Page 3-46. This Equation will be rearranged so as to accommodate the values and units provided in the problem statement; thus:

$$Q = AV = \frac{\pi D^2 V}{4} = \frac{\pi d^2 V}{576}$$

$$Q = \frac{\pi[15.0]^2[150][60]}{576} = \frac{\pi[225][9,000]}{576}$$

$$Q = \frac{6,361,725.1}{576} = 11,044.7$$

Now, with the Volumetric Flow Rate for the smaller system known, we can apply **Equation #26A** from Page 3-56 to determine the Volumetric Flow Rate in the new larger system; thus:

$$\frac{CFM_1}{CFM_2} = \left[\frac{D_1}{D_2}\right]^3$$

Transposing and solving for "CFM$_2$", we get:

$$CFM_2 = CFM_1\left[\frac{D_2}{D_1}\right]^3$$

$$CFM_2 = [11,044.7]\left[\frac{18.0}{15.0}\right]^3 = [11,044.7][1.20]^3$$

$$CFM_2 = [11,044.7][1.728] = 19,085.18$$

Now that we have the new Volumetric Flow Rate, it will be very easy to determine the new Duct Velocity. We will simply apply **Equation #23A** from Page 3-46, again rearranged to suit both the units and values provided in the problem statement; thus:

$$V = \frac{Q}{A} = \frac{576Q}{\pi d^2}$$

$$V = \frac{[576][19,085.18]}{\pi[18.0]^2} = \frac{10,993,061.0}{\pi[324]}$$

$$V = \frac{10,993,061.0}{1,017.88} = 10,800 \text{ fpm}$$

$$V_{fps} = \frac{10,800}{60} = 180 \text{ fps}$$

∴ The new Duct Velocity will become 180 fps = 10,800 fpm.

Problem #128:

To solve this problem, we must first determine the Volumetric Flow Rate in the 8.0 inch ducts. Again, to do this, we will simply apply **Equation #23C** from Page 3-47; thus:

$$V = 4,005 \sqrt{VP}$$

Transposing and solving for "VP", we get:

$$VP = \left[\frac{V}{4,005} \right]^2$$

$$VP = \left[\frac{7,525}{4,005} \right]^2 = [1.879]^2 = 3.530$$

Now, since we have successfully calculated the Velocity Pressure in the Duct, and since we were given the Total Pressure in the Duct, we can apply **Equation #23E** from Page 3-48 to determine the Static Discharge Pressure in the smaller sized (ie. 8.0 inch diameter) Duct and Fan.

$$TP_d = SP_d + VP_d$$

Transposing and solving for "SP_d", we get:

$$SP_d = TP_d - VP_d$$

$$SP_d = 9.8 - 3.53 = 6.27$$

Now we have the information that is required to calculate the requested Fan Static Discharge Pressure for the larger Fan System. To do so, we will make use of **Equation #26B** from Page 3-56; thus:

$$\frac{SP_{FAN_1}}{SP_{FAN_2}} = \left[\frac{D_1}{D_2} \right]^2$$

Transposing and solving for "SP_{FAN_2}", we get:

$$SP_{FAN_2} = SP_{FAN_1}\left[\frac{D_2}{D_1}\right]^2$$

$$SP_{FAN_2} = [6.27]\left[\frac{[8.0][1.50]}{8.0}\right]^2 = [6.27][1.50]^2$$

$$SP_{FAN_2} = [6.27][2.25] = 14.11$$

∴ The Fan Static Discharge Pressure for the New System will be 14.1 inches of water.

Problem #129:

The solution to this problem will rely on **Equation #26C** on Page 3-57, thus:

$$\frac{BHP_1}{BHP_2} = \left[\frac{D_1}{D_2}\right]^5$$

Transposing and solving for "D_2", we get:

$$D_2 = D_1\sqrt[5]{\frac{BHP_2}{BHP_1}}$$

$$D_2 = 9.0\sqrt[5]{\frac{10}{75}} = [9.0]\sqrt[5]{0.133}$$

$$D_2 = [9.0][0.668] = 6.02$$

∴ The system could be downsized to an overall 6.0 inch diameter configuration and fully satisfy all stated conditions.

Problem #130:

To solve this problem, we will have to use **Equation #27B** on Page 3-58; thus;

$$BHP = 0.0157 \left[\frac{[CFM][TP]}{FME} \right]$$

$$BHP = \frac{[0.0157][22,500][3.4]}{65}$$

$$BHP = \frac{1,201.05}{65} = 18.48$$

∴ The Brake Horsepower of the Subject Fan is approximately 18.5 bhp.

Problem #131:

To solve this problem, we must first determine the Volumetric Flow Rate for the system being evaluated. To do this, we will again rely on **Equation #23A** on Page 3-46. Once this has been done, we will be able to use **Equation #27A** on Page 3-58 to determine the required Fan Mechanical Efficiency.

$$Q = AV = \frac{\pi D^2 V}{4} = \frac{\pi d^2 V}{576}$$

$$Q = \frac{\pi [12.0]^2 [6,600]}{576} = \frac{\pi [144][6,600]}{576}$$

$$Q = \frac{2,985,769.7}{576} = 5,183.63$$

Now since we know the Volumetric Flow Rate, we can apply **Equation #27A** and calculate the required Fan Mechanical Efficiency; thus:

$$BHP = 0.0157 \left[\frac{[CFM][TP]}{FME} \right]$$

Transposing and solving for "FME", we get:

$$FME = 0.0157 \left| \frac{[CFM][TP]}{BHP} \right|$$

$$FME = [0.0157] \left| \frac{[5,183.6][15.0]}{20} \right|$$

$$FME = [0.0157] \left| \frac{77,754.42}{20} \right|$$

$$FME = [0.0157][3,887.72] = 61.04$$

∴ The Fan Mechanical Efficiency is 61%.

Problem #132:

Ultimately, the solution to this problem will use **Equation #27B** on Page 3-58; however, we must first determine the two Fan Total Pressure terms. To do this we will use the three relationships listed on Pages 3-47 and 3-48, identified as **Equation #'s 23D, 23E, & 23F.** We start by deriving the relationship for TP_{out}; thus:

$$TP_{out} = VP_{out} + losses$$

$$TP_{out} = 5.6 + 4.8 = 10.4 \text{ inches of water}$$

Next, we do the same for TP_{in}; thus:

$$TP_{in} = VP_{in} + losses$$

$$TP_{in} = 4.2 + 4.5 = 8.7 \text{ inches of water}$$

Note, here we are evaluating the Fan Total Input Pressure, namely a Total Pressure on the inlet side of the fan. Fan Total Input Pressures are <u>always</u> <u>negative</u>, since they are on the suction side of the Fan.

$$\therefore TP_{in} = -8.7 \text{ inches of water}$$

Now, finally using **Equation #27B**, we get:

$$TP_{FAN} = TP_{out} + TP_{in}$$

$$TP_{FAN} = 10.4 - [-8.7] = 10.4 + 8.7 = 19.1$$

> $\therefore$ The Fan Total Pressure in this case is 19.1 inches of water.

Problem #133:

Again, we will eventually determine the Fan Total Pressure by using **Equation #27B** from Page 3-58. To start with, however, we must again deal with the Fan Total Input Pressure term. In doing this, we will use a combination of **Equation # 23E** from Page 3-48 and Equation # 27B from Page 3-58; thus:

$$TP_{FAN} = TP_{out} - TP_{in} \quad [\#27B]$$

$$TP_{in} = SP_{in} + VP_{in} \quad [\#23E]$$

$$TP_{FAN} = TP_{out} - SP_{in} - VP_{in} \quad [\text{the combination}]$$

Rewriting **Equation #23E** in terms of the output side of the Fan, and substituting in the values we have been given for the Fan Total Output Pressure and the Velocity Pressure at the Fan outlet, we get:

$$TP_{out} = SP_{out} + VP_{out}$$

$$25.7 = SP_{out} + 13.4$$

$$SP_{out} = 25.7 - 13.4 = 12.3$$

We have been given that $\Delta SP = 19.6$ inches of water; and we know that:

$$\Delta SP = SP_{out} - SP_{in}$$

$$\therefore \ SP_{in} = SP_{out} - \Delta SP$$

$$SP_{in} = 12.3 - 19.6 = -7.3$$

Now, finally we have all the data to use the "combination" relationship developed above; thus:

$$TP_{FAN} = TP_{out} - SP_{in} - VP_{in}$$

$$TP_{FAN} = 25.7 - [-7.3] - 13.4 = 25.7 + 7.3 - 13.4 = 19.6$$

∴ The Fan Total Pressure in this case is 19.6 inches of water.

Problem #134:

To solve this problem, we must eventually apply **Equation #27C** from Page 3-59. First, however, we must determine the Static Pressure at the Fan outlet, and to do so, we must use **Equation #23E** from Page 3-48; thus:

$$TP_{out} = SP_{out} + VP_{out}$$

$$16.3 = SP_{out} + 8.2$$

$$SP_{out} = 16.3 - 8.2 = 8.1$$

We can now apply **Equation #27C**; thus:

$$SP_{FAN} = SP_{out} - SP_{in} - VP_{in}$$

$$SP_{FAN} = 8.1 - [-8.3] - 6.4 = 8.1 + 8.3 - 6.4 = 10.0$$

∴ The Fan Static Discharge Pressure = 10.0 inches of water.

Problem #135:

To solve this problem, we need only apply **Equation #27C** from Page 3-59; thus:

$$SP_{FAN} = SP_{out} - SP_{in} - VP_{in}$$

$$21.5 = 14.4 - [-12.7] - VP_{in} \quad \&$$

$$VP_{in} = 14.4 + 12.7 - 21.5 = 5.6$$

∴ The Velocity Pressure at this Fan's inlet is 5.6 inches of water.

Problem #136:

The solution to this problem relies on the relationships listed under **Equation #28A** on Page 3-60 — the Duct Junction Balancing System. This problem asks us to balance a somewhat complex set of Duct Junctions, involving three separate Ducts that join to form one single Duct. Let us begin by examining two of these Duct Junctions, namely: [AC - DD] & [CD - DD]. Specifically, let us examine the ratio of the Static Pressures listed for these two (these two Duct Junctions were chosen simply because the <u>absolute magnitude</u> of the <u>difference</u> in their listed Static Pressures was the <u>greatest</u> in this group of three Duct Junctions — ie. the Absolute Value of the difference of their listed Static Pressures = $|\Delta SP| = |SP_{greater} - SP_{lesser}| = |1.44 - 1.26| = 0.18$ inches of water); thus, using **Equation #28A** as stated earlier, we get:

$$R = \frac{SP_{greater}}{SP_{lesser}}$$

$$R = \frac{1.44}{1.26} = 1.143$$

For any pair of Duct Junctions having this ratio: $1.05 \leq R \leq 1.20$, in order to achieve an adequate balance, we must <u>increase</u> the Flow Volume in the branch having the <u>lesser</u> Flow Volume, according to the relationship listed under the second of the three Ratio Scenarios that have been tabulated under **Equation #28A**; thus:

$$Q_{new} = Q_{former} \sqrt{\frac{SP_{greater}}{SP_{lesser}}}$$

$$Q_{new} = 520 \sqrt{\frac{1.44}{1.26}} = 520\sqrt{1.143}$$

$$Q_{new} = [520][1.069] = 555.90 \sim 556$$

> ∴ To balance the [AC - DD] & [CD - DD] Duct Junctions, we must increase the Flow Volume in Duct AD from its current value of 520 cfm, to ~ 556 cfm.

Let us now address Duct Junctions, [BD - DD] & [CD - DD], in the same manner; thus:

$$R = \frac{SP_{greater}}{SP_{lesser}}$$

$$R = \frac{1.44}{1.39} = 1.036$$

For this pair of Duct Junctions, which pair has as its ratio: R < 1.05, we can consider that this Junction is in balance.

> ∴ We can consider this pair of Duct Junctions to be in balance; thus, we should do nothing to the Flow Volumes in either of them.

Problem #137:

To solve the final part of this problem, we will have to apply <u>both</u> Dalton's Law of Partial Pressures — as specified by **Equation #7B** on Page 3-17 — and Raoult's Law — as specified by **Equation #7C** on Page 3-18 — to the solution obtained in the first part of the problem. To solve the first part of this problem, we must use **Equation #29A** on Page 3-61; thus:

$$C = \left| \frac{V_s \rho T}{[MW_s] P_{atm} V_{room}} \right| [6.24 \times 10^7]$$

Before substituting values into this somewhat complex formula, we must determine that we are using the units that this relationship calls for. We note that:

(1) the Solvent Volume term, "V_s", must be in milliliters [thus, we must <u>multiply</u> the Solvent Volume that is provided in the problem data by a factor of 1,000 ml/liter];

(2) the Temperature term, "T", must be in °K (thus, we must <u>add</u> 273.16° to the Celsius Temperature that is provided in the problem data); and

(3) the Room Volume term, "V_{room}", must be in liters (thus, we must multiply the Room Volume that is provided in the problem data by a factor of 28.32 liters/ft^3).

When we make these substitutions, we get:

$$C = \left| \frac{(1.5)(1,000)(0.791)(273.16 + 25)}{(58.08)(760)(1,650)(28.32)} \right| \left[6.24 \times 10^7 \right]$$

$$C = \frac{[1.5][1,000][0.791][298.16]\left[6.24 \times 10^7\right]}{[58.08][760][1,650][28.32]}$$

$$C = \frac{2.208 \times 10^{13}}{2.063 \times 10^9} = 10,702.48$$

$\therefore$ The Ultimate Ambient Concentration Level of the Acetone Vapors in this Room will be ~ 10,702 ppm.

According to Raoult's Law, the Partial Vapor Pressure of any volatile component will be the product of the Vapor Pressure of the pure component and the Mole Fraction of that component in the solution being considered. In this case we are dealing with pure acetone; thus, the Mole Fraction will be 1.00 = 100%, and from Raoult's Law we see that the "potential" for the Partial Vapor Pressure of acetone would simply be its pure state vapor pressure, or 226 mm Hg. Applying Dalton's Law of Partial Pressures to this fact, we get:

$$C_{max} = \frac{[1,000,000]\left[PVP_{acetone}\right]}{P_{ambient}}$$

$$C_{max} = \frac{[1,000,000][226]}{760} = 297,368.4$$

$\therefore$ It is reasonable to assume that <u>all</u> the Acetone will evaporate, since the Concentration that would be produced by the Quantity of acetone in the flask that was broken during the earthquake is only a <u>small</u> <u>fraction</u> of the Theoretical Maximum Acetone Concentration that could exist in the Ambient Air (10,702 ppm vs. 297,368 ppm, with the former being only 3.6% of the latter). In fact, in a room of the size given in this problem, 41.7 liters of pure Acetone could reasonably be expected to evaporate completely!

Problem #138:

To solve this extension to the previous problem, we must apply **Equation #29B** from Page 3-62; thus:

$$D_t = \left[\frac{V}{Q}\right] ln \left[\frac{C_{initial}}{C_{ending}}\right]$$

$$D_t = \left[\frac{1,650}{500}\right] ln \left[\frac{10,702.48}{750}\right]$$

$$D_t = 3.30 \, ln \, 14.27$$

$$D_t = [3.30][2.658] = 8.772$$

∴ It will take ~ 8.77 minutes to ventilate the room sufficiently well to have achieved the desired 750 ppm ambient concentration level of acetone.

Problem #139:

The solution to this problem also will eventually employ **Equation #29A** from Page 3-61; however, first we must determine what the maximum ambient concentration of ethanol could be at NTP. To do this, we will again have to use Dalton's Law of Partial Pressures, listed as **Equation #7B** on Page 3-17; thus:

$$C_{max} = \frac{[1,000,000][PVP_{ethanol}]}{P_{ambient}}$$

$$C_{max} = \frac{[1,000,000][44.3]}{760} = 58,289.5$$

This means that ethanol will evaporate from his batch of "White Lightning" until the concentration in this cylindrical redwood tank/storeroom has attained a value of ~ 58,290 ppm. Knowing this, we can begin to work with **Equation #29A**; thus:

$$C = \left[\frac{V_s \rho T}{[MW_s] P_{atm} V_{room}}\right][6.24 \times 10^7]$$

Again, before substituting values into this somewhat complex formula, we must determine that we are using the units that this relationship calls for. In particular, we must note that:

(1) the Solvent Volume term, "V_s", must be in milliliters [thus, we must <u>multiply</u> the Solvent Volume that is provided in the problem data by a factor of 3,785.4 ml/gallon]; and

(2) the Temperature term, "T", must be in °K (thus, we must <u>add</u> 273.16° to the Celsius Temperature that is provided in the problem data).

Before making these substitutions, we should transpose the given Equation, and solve for the Room Volume, "V_{room}", which in this case will be the volume of the cylindrical redwood tank/storeroom. When we have done this, and have made these substitutions, we get:

$$V_{room} = \left| \frac{V_s \rho T}{[MW_s]P_{atm}C} \right| [6.24 \times 10^7]$$

$$V_{room} = \left| \frac{(1)(3,785.4)(0.789)(273.16+25)}{(46.07)(760)(58,289.5)} \right| [6.24 \times 10^7]$$

$$V_{room} = \frac{[3,785.4][0.789][298.16][6.24 \times 10^7]}{[46.07][760][58,289.5]}$$

$$V_{room} = \frac{5.557 \times 10^{13}}{2.040 \times 10^9} = 27,227.06$$

We now obviously have the volume of this cylindrical redwood tank/storeroom in liters (namely: ~ 27,277 liters); however, we must convert this volume into cubic feet in order to obtain the requested result in the most direct manner possible.

$$V_{cubic\ feet} = \frac{27,227.06}{28.32} = 961.41$$

Now we must express the desired redwood tank/storeroom diameter, "d", in terms that will permit us to use this volume, "$V_{storeroom}$", and the tank height, "h" (which has been given as being 6 feet) to advantage, as we proceed in obtaining the requested result; thus:

$$V_{storeroom} = \frac{\pi d^2 h}{4}$$

$$d^2 = \frac{4 V_{storeroom}}{\pi h}$$

$$d = \sqrt{\frac{4 V_{storeroom}}{\pi h}} = 2\sqrt{\frac{V_{storeroom}}{\pi h}}$$

Now, all we have to do is to substitute in the known values to obtain the requested diameter; thus:

$$d = 2\sqrt{\frac{961.41}{6\pi}} = 2\sqrt{51.004}$$

$$d = [2][7.142] = 14.283$$

∴ The diameter of this Moonshiner's cylindrical redwood tank/storeroom is 14.3 feet.

Problem #140:

To solve this problem, we must use **Equation #29B** from Page 3-62; thus:

$$D_t = \left[\frac{V}{Q}\right] ln \left\lfloor \frac{C_{initial}}{C_{ending}} \right\rfloor$$

$$D_t = \frac{961.41}{25} ln \left\lfloor \frac{59,289.5}{1,000} \right\rfloor$$

$$D_t = 38.46 ln \, 58.29$$

$$D_t = [38.46][4.065] = 156.34$$

∴ He will have to run his ventilation fan for 156.3 minutes, or 2 hours and 36+ minutes in order to achieve his maximum ambient ethanol concentration goal of 1,000 ppm.

Problem #141:

To solve this problem we must apply **Equation #29C** on Page 3-63, thus:

$$C = C_0 e^{-\left[V_{removed}/V_{room}\right]}$$

For this problem, we have been asked to determine the number of *Room Volumes* that must be removed from some identifiable space, in order to achieve some well defined and specific decrease in the "starting ambient concentration" of some unidentified volatile substance. From the perspective of the applicable formula listed above, we must view this as asking for a value of "n", where "n" is the number of *Room Volumes* for which — once this volume of ambient, volatile filled air had been removed from the space — would result in a situation where the residual room concentration of that volatile would be at or below the identified target concentration level. Specifically, we are seeking an exponent of "e" in the following generalized format:

$$-\frac{n\,V_{room}}{V_{room}}$$

Clearly, the "V_{room}" terms will cancel out, and we are left with the simple exponent value of "n"; and we can, therefore, see that the formula evolves to the following:

$$C = C_0 e^{-\frac{n\,V_{room}}{V_{room}}} = C_0 e^{-n}$$

The task in this problem, then, is simply to determine the value of "n", as a number of *Room Volumes*, that corresponds to:

(1) a decrease in the ambient concentration to a level that is only 10% of the starting value [ie. the ending concentration, "$C_{90\%}$", has the value $0.1C_0$]; and

(2) a decrease in the ambient concentration to a level that is only 1% of the starting value [ie. the ending concentration, "$C_{99\%}$", has a value, $0.01C_0$].

Let us consider these two situations in order; thus:

1. For a 90% reduction in the concentration: $C_{90\%} = 0.1C_0 = C_0 e^{-n_{90\%}}$

$$0.1 = e^{-n_{90\%}}$$

Now we must take the natural logarithm of both sides of the equation, thus:

$$ln\,0.1 = -n_{90\%} = -2.303$$

$$n_{90\%} = 2.303$$

2. For a 99% reduction in the concentration: $\quad C_{99\%} = 0.01\,C_0 = C_0\,e^{-n_{99\%}}$

$$0.01 = e^{-n_{99\%}}$$

Again, taking the natural logarithm of both sides:

$$ln\,0.01 = -n_{99\%} = -4.605$$

$$n_{99\%} = 4.605$$

∴ To achieve specified reductions in the ambient concentrations of <u>any</u> volatile substance, one must purge the following number of *Room Volumes* to attain the identified target reduction in the ambient room concentration level:

<u>Target Reduction, as a %</u>	# of *Room Volumes*
90%	~ 2.3
99%	~ 4.6

Problem #142:

The solution to this problem can be determined from the several relationships listed under the statistical parameters that are applicable to a Normal Distribution. Each will be identified as it is used.

The Reader is advised that, each of the mathematical operations that are used to determine one of the parameters for which the Problem Statement has asked will rely on, and be carried out by using, a calculator that has a <u>full library</u> of statistical functions. As a result of this, there will likely be some differences between the results listed in the various parts of this Problem, and those that a Reader, who chooses to follow the calculation process stepwise from start to finish, will obtain. Most calculators having statistical capabilities will maintain intermediate results in their internal memory banks, carrying each result to <u>many more decimal places</u> than would be practical for a hand calculating process. Because of this, step-by-step hand calculations that use numbers accurate to only two or three decimal places, will likely ob-

tain slightly different results than are obtained by the more precise statistical sub-
routines in a calculator that posses statistical capabilities.

At this point, it will be prudent to reorganize the data provided in this Problem into
a tabulation of several useful columns; thus:

1 Dept.	2 x_i	3 $log\, x_i$	4 $x_i - \mu$	5 $[x_i - \mu]^2$
4	43	1.633	- 34.40	1,183.36
7	55	1.740	- 22.40	501.76
10	62	1.792	- 15.40	237.16
12	62	1.792	- 15.40	237.16
9	63	1.799	- 14.40	207.36
15	69	1.839	- 8.40	70.56
2	71	1.851	- 6.40	40.96
11	77	1.886	- 0.40	0.16
14	82	1.914	4.60	21.16
1	85	1.929	7.60	57.76
6	87	1.940	9.60	92.16
5	90	1.954	12.60	158.76
13	95	1.978	17.60	309.76
3	102	2.009	24.60	605.16
8	118	2.072	40.60	1,648.36

We must now summarize the second, third, and fifth columns to obtain values that
will be useful in several subsequent calculations, thus:

For Column #2 — x_i:
$$\sum_{i=1}^{15} x_i = 1,161$$

For Column #3 — $log\, x_i$:
$$\sum_{i=1}^{15} log\, x_i = 28.130$$

For Column #5 — $[x_i - \mu]^2$:
$$\sum_{i=1}^{15} [x_i - \mu]^2 = 5,371.6$$

SOLUTIONS

We can now determine all the required statistical parameters using the appropriate relationship; thus:

The Mean $= \mu = \overline{x}$

The Mean, "μ" or "$\overline{x}$", is given by **Equation #30A** on Page 3-64; thus:

$$\mu = \overline{x} = \frac{1}{n} \sum_{i=1}^{n} x_i$$

$$\mu = \overline{x} = \frac{1}{15} \sum_{i=1}^{15} x_i = \frac{1,161}{15} = 77.40$$

$\therefore$ The Mean = 77.40 days

The Median $= m_e$

The Median, "m_e", is the "Midpoint" Value of the Distribution, and is given by **Equation #30B** on Page 3-64. For this Distribution, the Median is the number of workdays without a loss time accident that is associated with Department #11. For this value, 77 days, there are 7 larger values and 7 smaller ones; thus:

$\therefore$ The Median = 77 days

The Mode $= m_0$

The Mode, "m_0", is the "Most Populous" Value in the Distribution, and is given by **Equation #30C** on Page 3-65. For this Distribution, the Mode is the number of workdays without a loss time accident that is associated with Department #'s 10 & 12. This is, in fact, the only tabulated value for which there is more than a single entry. For this one, 62 days, there are two entries; these are associated, as stated above, with Department #'s 10 & 12; thus:

$\therefore$ The Mode = 62 days

The Range = R

The Range, "R", of any Distribution is given by **Equation #30D** on Page 3-65, thus:

$$R = \left[x_{i_{max}} - x_{i_{min}} \right]$$

$$R = 118 - 43 = 75$$

$\therefore$ The Range = 75 days

The Geometric Mean = $M_{geometric}$

The Geometric Mean, "$M_{geometric}$", is given by **Equation #30E** on Page 3-66; thus:

$$M_{geometric} = 10^{\frac{1}{n}\sum_{i=1}^{n} log x_i}$$

$$M_{geometric} = 10^{28130/15} = 10^{1.875} = 75.04$$

$\therefore$ The Geometric Mean = 75.0 days

The Sample Variance = s^2

The Sample Variance, "s^2", is given by **Equation #31A** on Page 3-67; thus:

$$s^2 = \frac{\sum_{i=1}^{n} \left[x_i - \mu \right]^2}{n - 1}$$

$$s^2 = \frac{5,371.6}{15 - 1} = \frac{5,371.6}{14} = 383.69$$

$\therefore$ The Sample Variance = 383.7 days²

The Population Variance = σ^2

The Population Variance, "σ^2", is given by **Equation #31B** on Page 3-68; thus:

$$\sigma^2 = \frac{\sum_{i=1}^{n}[x_i - \mu]^2}{n}$$

$$\sigma^2 = \frac{5,371.6}{15} = 358.11$$

∴ The Population Variance = 358.11 days²

The Sample Standard Deviation = s

The Sample Standard Deviation, "s", is given by **Equation #31C** on Page 3-69; thus:

$$s = \sqrt{\frac{\sum_{i=1}^{n}[x_i - \mu]^2}{n-1}}$$

$$s = \sqrt{\frac{5,371.6}{15-1}} = \sqrt{\frac{5,371.6}{14}}$$

$$s = \sqrt{383.686} = 19.59$$

∴ The Sample Standard Deviation = 19.6 days

The Population Standard Deviation = σ

The Population Standard Deviation, "σ", is given by **Equation #31D** on Page 3-70; thus:

$$\sigma = \sqrt{\frac{\sum\limits_{i=1}^{n}\left[x_i - \mu\right]^2}{n}}$$

$$\sigma = \sqrt{\frac{5,371.6}{15}} = \sqrt{358.107} = 18.92$$

∴ The Population Standard Deviation = 18.9 days

The Sample Coefficient of Variation = CV_{sample}

The Sample Coefficient of Variation, "CV_{sample}", is given by **Equation #32A** on Page 3-71; thus:

$$CV_{sample} = \frac{s}{\mu}$$

$$CV_{sample} = \frac{19.588}{77.40} = 0.253$$

∴ The Sample Coefficient of Variation = 0.253

The Population Coefficient of Variation = $CV_{population}$

The Population Coefficient of Variation, "$CV_{population}$", is given by **Equation #32B** on Page 3-71; thus:

$$CV_{population} = \frac{\sigma}{\mu}$$

$$CV_{population} = \frac{18.924}{77.40} = 0.244$$

∴ The Population Coefficient of Variation = 0.244